WORKBOOKS IN CHEMISTRY
ORGANIC CHEMISTRY

Periodic table of the elements

Key: Atomic number (8), Symbol (O), Relative atomic mass (15.999)

Group	1	2		3	4	5	6	7	8	9	10	11	12	13	14	15	16	17	18
Period 1	1 H 1.0079																		2 He 4.0026
2	3 Li 6.941	4 Be 9.0122												5 B 10.811	6 C 12.011	7 N 14.007	8 O 15.999	9 F 18.998	10 Ne 20.180
3	11 Na 22.990	12 Mg 24.305												13 Al 26.982	14 Si 28.086	15 P 30.974	16 S 32.065	17 Cl 35.453	18 Ar 39.948
4	19 K 39.098	20 Ca 40.078		21 Sc 44.956	22 Ti 47.867	23 V 50.942	24 Cr 51.996	25 Mn 54.938	26 Fe 55.845	27 Co 58.933	28 Ni 58.693	29 Cu 63.546	30 Zn 65.409	31 Ga 69.723	32 Ge 72.64	33 As 74.922	34 Se 78.96	35 Br 79.904	36 Kr 83.798
5	37 Rb 85.468	38 Sr 87.62		39 Y 88.906	40 Zr 91.224	41 Nb 92.906	42 Mo 95.94	43 Tc (98)	44 Ru 101.07	45 Rh 102.91	46 Pd 106.42	47 Ag 107.87	48 Cd 112.41	49 In 114.82	50 Sn 118.71	51 Sb 121.76	52 Te 127.60	53 I 126.90	54 Xe 131.29
6	55 Cs 132.91	56 Ba 137.33		57 La 138.91	72 Hf 178.49	73 Ta 180.95	74 W 183.84	75 Re 186.21	76 Os 190.23	77 Ir 192.22	78 Pt 195.08	79 Au 196.97	80 Hg 200.59	81 Tl 204.38	82 Pb 207.2	83 Bi 208.98	84 Po (209)	85 At (210)	86 Rn (222)
7	87 Fr (223)	88 Ra (226)		89 Ac (227)	104 Rf (263)	105 Db (262)	106 Sg (266)	107 Bh (272)	108 Hs (277)	109 Mt (276)	110 Ds (281)	111 Rg (280)	112 Cn (277)	113 Nh unknown	114 Fl (289)	115 Mc unknown	116 Lv (298)	117 Ts unknown	118 Og unknown

s-block | *d-block* | *p-block*

Lanthanides (Period 6):

58 Ce 140.12	59 Pr 140.91	60 Nd 144.24	61 Pm (145)	62 Sm 150.36	63 Eu 151.96	64 Gd 157.25	65 Tb 158.93	66 Dy 162.50	67 Ho 164.93	68 Er 167.26	69 Tm 168.93	70 Yb 173.04	71 Lu 174.97

Actinides (Period 7):

90 Th 232.04	91 Pa 231.04	92 U 238.03	93 Np (237)	94 Pu (244)	95 Am (243)	96 Cm (247)	97 Bk (247)	98 Cf (251)	99 Es (252)	100 Fm (257)	101 Md (258)	102 No (259)	103 Lr (262)

f-block

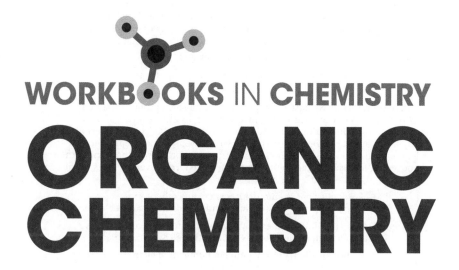

WORKBOOKS IN CHEMISTRY
ORGANIC CHEMISTRY

Michael Cook
University of Hertfordshire

Philippa Cranwell
University of Reading

Series editor
Elizabeth Page
University of Reading

OXFORD
UNIVERSITY PRESS

Great Clarendon Street, Oxford, OX2 6DP,
United Kingdom

Oxford University Press is a department of the University of Oxford.
It furthers the University's objective of excellence in research, scholarship,
and education by publishing worldwide. Oxford is a registered trade mark of
Oxford University Press in the UK and in certain other countries

© M Cook & P Cranwell 2017

The moral rights of the authors have been asserted

Impression: 1

All rights reserved. No part of this publication may be reproduced, stored in
a retrieval system, or transmitted, in any form or by any means, without the
prior permission in writing of Oxford University Press, or as expressly permitted
by law, by licence or under terms agreed with the appropriate reprographics
rights organization. Enquiries concerning reproduction outside the scope of the
above should be sent to the Rights Department, Oxford University Press, at the
address above

You must not circulate this work in any other form
and you must impose this same condition on any acquirer

Published in the United States of America by Oxford University Press
198 Madison Avenue, New York, NY 10016, United States of America

British Library Cataloguing in Publication Data

Data available

ISBN 978-0-19-872951-8

Printed in Great Britain by
Bell & Bain Ltd., Glasgow

Links to third party websites are provided by Oxford in good faith and
for information only. Oxford disclaims any responsibility for the materials
contained in any third party website referenced in this work.

Overview of contents

Preface vii

1 Foundations

1.1	Drawing organic structures	1
1.2	Naming organic structures	4
1.3	Orbital overlap and bonding	8
1.4	Orbital hybridization	11
1.5	Double bond equivalents	15
1.6	Polarity	18
1.7	Aromaticity	21
1.8	Resonance	24
1.9	Tautomerism	28
1.10	Synoptic questions	30

2 Isomerism

2.1	What is isomerism?	32
2.2	Constitutional isomerism	32
2.3	Configurational isomerism	35
2.4	*Cis/trans* isomerism	36
2.5	Optical isomerism (chirality)	39
2.6	Synoptic questions	45

3 Nucleophilic substitution

3.1	Electrophiles and nucleophiles	47
3.2	Lewis acids and bases	49
3.3	S_N1 and S_N2	51
3.4	The impact of pK_a on leaving group ability	56
3.5	Synoptic questions	61

4 Elimination reactions

4.1	Synthesis of alkene via elimination (E2, E1, and E1cB)	62

5 Reactions of unsaturated compounds

5.1	Electrophilic addition	67

6 Aromatic chemistry

6.1	Electrophilic aromatic substitution	72
6.2	Effects of directing groups on S_EAr	74
6.3	Nucleophilic aromatic substitution	79
6.4	Azo coupling	82
6.5	Synoptic questions	85

7 Carbonyl chemistry

7.1	Structure and bonding	87
7.2	Reactions with nucleophiles	91
7.3	Reactions with reducing agents	94
7.4	Carboxylic acids	97
7.5	Acyl chlorides	100
7.6	Esters	103
7.7	Amides	106
7.8	Synoptic questions	109

Synoptic questions   111
Answers   113
Appendix 1 Acidity constants   115
Appendix 2 Electronegativity values for common elements   117
Appendix 3 Common functional groups in decreasing order of seniority, according to IUPAC   118
Index   119

Preface

Welcome to the Workbooks in Chemistry

The Workbooks in Chemistry have been designed to offer additional support to help you make the transition from school to university-level chemistry. They will also be useful if you are studying for related degrees, such as biochemistry, food science, or pharmacy.

Introduction to the Workbooks

The Workbooks cover the three traditional areas of chemistry: inorganic, organic and physical. They are designed to complement your first year chemistry modules and to supplement, but not replace, your course text book and lecture notes. You may want to use the Workbooks as self-test guides as you carry out a specific topic, or you may find them useful when you have finished a topic as you prepare for end of semester tests and exams. When preparing for tests and exams, students often use practice questions, but model answers are not always available. This is because there is usually more than one correct way to answer a question and your lecturers will want to give you credit for your problem-solving approach and working, as well as having obtained the correct answer. These Workbooks will give you guidance on good practice and a logical approach to problem solving, with plenty of hints and tips on how to avoid typical pitfalls.

Structure of the Workbooks

Each of the three Workbooks is divided into chapters covering the different topics that appear in typical first year chemistry courses. As external examiners and assessors at different UK and international universities, we realize that every chemistry programme is slightly different, so you may find that some topics are covered in more depth than you require, or that there are topics missing from your particular course. If this is the case, we would be interested in hearing your views! However, we are confident that the topics covered are representative, and that most first year students will meet them at some point.

Each chapter is divided into sections, and each section starts with a brief introduction to the theory behind the concepts to put the subsequent problems in context. If you need to, you should refer to your lecture notes and text books at this point to fully revise the theory.

Following the outline introduction to each topic, there are a series of **worked examples**, which are typical of the problems you might be asked to solve in workshops or exams. These examples contain fully worked solutions that are designed to give you the scaffolding upon which to base any future answers, and sometimes provide you with hints about how to approach these types of question and how to avoid common errors.

After the worked examples relating to a topic, you will find further **questions** of a similar type for you to practise. The numerical or 'short' answers to these problems can be found at the end of the book, whilst fully worked solutions are available on the Online Resource Centre. At the end of each book is a bank of **synoptic questions**, also with worked solutions on the Online Resource Centre. Synoptic questions encourage you to draw on concepts from multiple topics, helping you to use your broader chemical knowledge to solve problems.

You can find the series website at www.oxfordtextbooks.co.uk/orc/chemworkbooks.

How to use the Workbooks

You will probably refer to these Workbooks at different times during your first year course, but we envisage they will be most useful when preparing for examinations after you have done some initial revision.

It is a good idea to use the introductions to the topics to check your understanding and refresh your memory. The next step is to follow through the worked examples, or try them out yourself.

The **hints** will give you guidance on how to tackle the problem—for example, reminding you of points you may need to use from different areas of chemistry.

> ▶ **Hint** If you find it difficult to rotate the molecule so that the lowest priority group is facing away from you, then leave is where it is, assign the stereochemical configuration, and reverse the answer at the end (i.e. (R) goes to (S), (S) goes to (R))—you'll end up with the correct isomer that way!

The **comments** will typically relate to the worked solutions and might explain why a unit conversion has been used, for example, or give some background explanation for the maths used in the solution. The comments are designed to help you avoid the typical mistakes students make when approaching each particular type of problem. It is to be hoped that by being aware of these pitfalls you will be able to overcome them.

> ⮕ Note that inverting all the stereocenters in a chiral molecule gives the enantiomer, unless the compound is meso. By contrast, inverting some, but not all, of the stereocenters gives a diastereomer.

When you are happy you have mastered the worked examples, try the questions. To check your answers go to the back of the book, and to check your working look for the fully worked solutions at www.oxfordtextbooks.co.uk/orc/chemworkbooks.

The synoptic questions can be used as a final revision tool when you are confident with your understanding of the individual topics and want some final practice before the exam or test. Again, you will find answers at the back of the book and full solutions online.

Final comments

We hope you find these workbooks helpful in reinforcing your understanding of key concepts in chemistry and providing tips and techniques that will stay with you for the rest of your chemistry degree course. If you have any feedback on the Workbooks—such as aspects you found particularly helpful or areas you felt were missing—please get in touch with us via the Online Resource Centre. Go to www.oxfordtextbooks.co.uk/orc/chemworkbooks.

1 Foundations

1.1 Drawing organic structures

Organic chemistry involves the study of molecules containing carbon, which make up the majority of biological matter. Due to the ability of carbon to bond so that it forms long chains, the complexity of organic molecules can be very challenging. You may be familiar with the condensed formula or structural formulae often used to display organic molecules in introductory chemistry courses. However, in specialized organic chemistry we will almost always use skeletal formulae in order to quickly represent complex structures in a simple manner, and you will need to be able to understand what these mean.

To draw a skeletal structure:

- All hydrogen atoms that are bonded to carbon are not drawn—there are simply too many to bother!
- Chains of carbon atoms are simply drawn as zig-zags in which each connecting point represents a carbon atom, with bonded hydrogen atoms.
- Double and triple bonds are represented as two or three lines between connecting points, respectively.
- 'Heteroatoms', such as N, O, P, S, and the halogens, are written using their atomic symbol, as normal.

→ Note: you may hear the word 'catenation' used to describe the way that carbon atoms form chains.

→ Methods of representing propan-2-ol, also known as 'isopropyl alcohol':
  Molecular formula: C_3H_8O;
  Condensed formula: $(CH_3)_2CHOH$;
  Structural formula:

Skeletal formula:

Worked example 1.1A

Convert the structural formula of 4-aminobutan-2-one into the skeletal formula.

4-aminobutan-2-one

→ This 'zig-zagging' comes from the 109.5° bond angle found in tetrahedral carbon atoms. The bond angles for carbon atoms with double or triple bonds are slightly different—for more information, see the section on orbital hybridization (section 1.4).

→ You may sometimes be asked to draw Lewis structures. These are the same as skeletal structures, but include dots to represent lone pairs of electrons.

skeletal structure   **Lewis structure**

Solution

First, we do not need to draw any of the hydrogen atoms that are bonded to carbon. We will leave the hydrogen atoms of the amine (—NH_2), as this is a functional group and the hydrogen atoms are more likely to be important to its reactivity. Now, we will redraw the carbon chain as zig-zags, with each 'point' representing a carbon atom, while leaving the O and N heteroatoms labelled. The skeletal structure should now be complete.

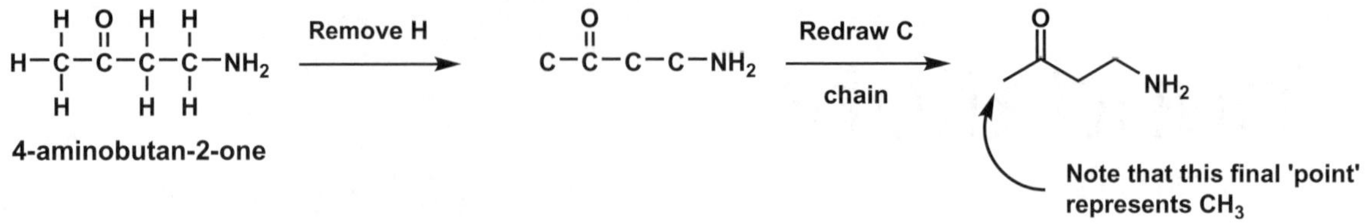

4-aminobutan-2-one

Note that this final 'point' represents CH$_3$

Worked example 1.1B

Methyl *tert*-butyl ether, or MTBE, is an organic solvent that is sometimes used as an alternative to diethyl ether, and also as an additive to unleaded petrol. Its condensed formula is CH$_3$OC(CH$_3$)$_3$. Draw out its structural and skeletal formulae.

Solution

You will not often be asked to draw the structural formulae of organic compounds, but it is useful here as an intermediate step between the molecular and skeletal formulae. Working from left to right along the carbon chain, drawing bonds at right-angles, we will arrive at the structural formula shown below. Care must be taken to ensure that the three methyl groups (CH$_3$) are drawn attached to the same carbon atom. From this point we can convert to the skeletal formula using the method in Worked example 1.1A. Note that the *tert*-butyl group, C(CH$_3$)$_3$, can be represented flat on the page, or in '3D'. Both options are shown, but representing skeletal structures in 3D will be very important when covering later topics.

➲ **Dashes and Wedges**: In order to represent 3D structures on the page, 'dashes' and 'wedges' can be used to show that a bond is pointing away from, or towards the viewer, respectively.

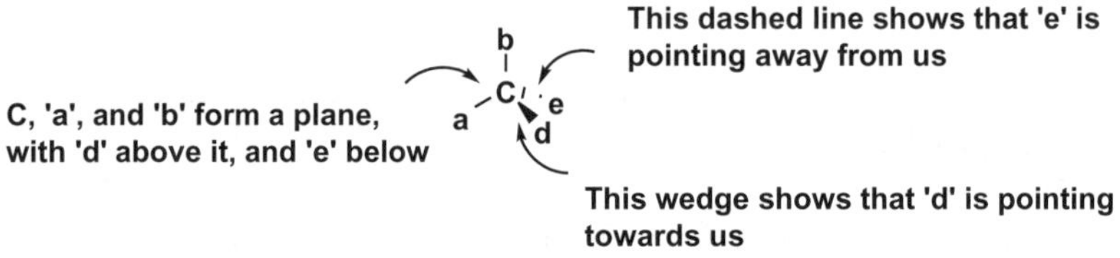

C, 'a', and 'b' form a plane, with 'd' above it, and 'e' below

This dashed line shows that 'e' is pointing away from us

This wedge shows that 'd' is pointing towards us

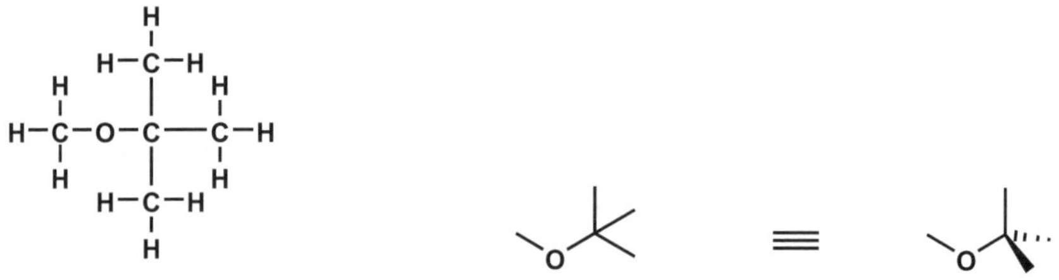

structural formula **'Flat' skeletal formula** ≡ **'3D' skeletal formula**

We can see from this example the level of complexity involved in drawing the full structural formula of a relatively simple molecule. Drawing the molecule as a skeletal formula reduces the image to a simple representation, while losing no structural information.

1.1 DRAWING ORGANIC STRUCTURES

Question 1.1

Draw the structural formulae of the following compounds from their condensed formulae.

(a) $CH_3CH_2CH_2CH(CH_3)_2$

(b) CH_2CHCH_2OH

(c) $CH_3CH_2CCCH_3$

(d) $(CH_3)_3CCH_2CH_2CH_2OH$

▶ **Hint** Be sure to look at the number of hydrogen atoms each carbon is bonded to—it may give you an idea as to whether a double or triple bond might be present.

Question 1.2

Convert the following structural formulae into skeletal formulae.

(a) but-2-ene

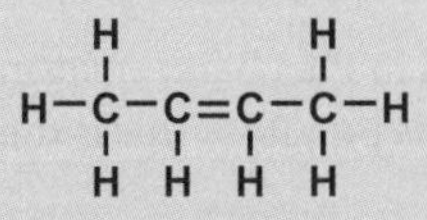

(b) 2-methoxybutane

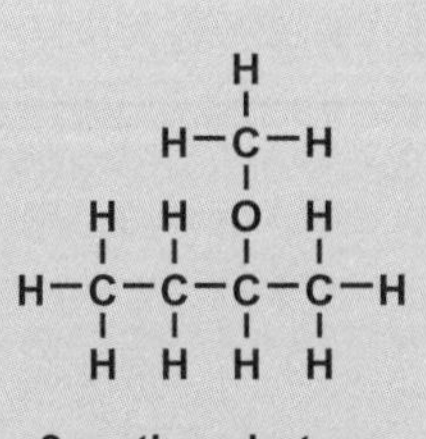

(c) butane-1,4-diamine (or 'putrescine')

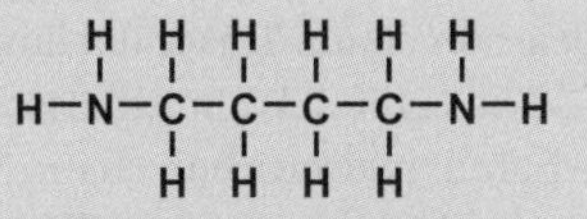

(d) cyclohexa-1,3,5-triene (or 'benzene')

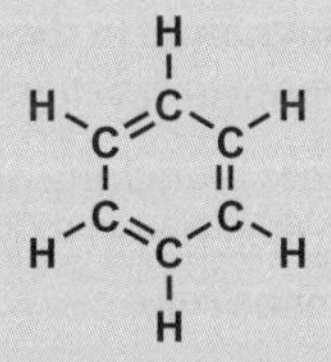

(e) 2-hydroxyacetonitrile (or 'glycolonitrile')

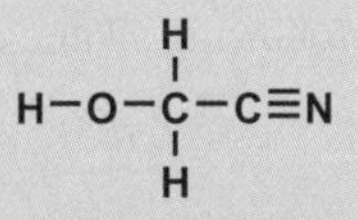

(f) ethyl butyrate (smells like pineapple!)

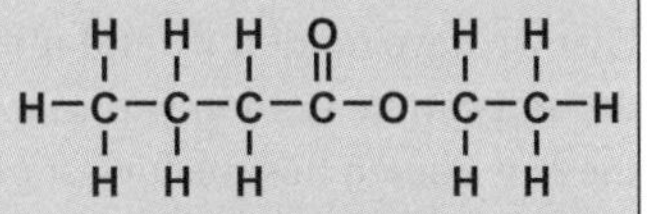

▶ **Hint** Take care when looking at double bonds—ensure that you draw the substituents in the correct positions relative to each other. This will crop up in stereochemistry, which is covered in Chapter 2.

➔ Remember to take care with the bond angles on alkenes and alkynes; C=C bonds give 120° angles with their substituents, while C≡C give bond angles of 180°. The reasons for this are explained in section 1.4.

→ There are a number of 'trivial' names which are still used by chemists today, but don't follow IUPAC rules. You will come across these names often, and include: acetone, formaldehyde, toluene, etc. Whilst these are useful for simplifying the naming of molecules, there is no consistent methodology used so they need to be learnt individually.

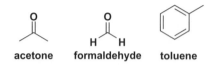

acetone   formaldehyde   toluene

→ Please note that when the numbering is conducted in this manner, you will sometimes see the positions described as 'locants'.

→ Please note, if there is more than one identical substituent, the prefixes di-, tri-, tetra-, etc., are used—this additional prefix does not affect the alphabetical ordering of multiple substituents. For example ethyl- would be placed before dimethyl- in an IUPAC name.

1.2 Naming organic structures

In order to be able to know the structure of a molecule from its name, a set of rules for naming organic compounds, 'nomenclature', has been laid out by the International Union of Pure and Applied Chemistry (IUPAC). This means that if you know the IUPAC name of a molecule, you should be able to draw it. This is a great idea in principle, but can get complicated very easily. In this workbook we will introduce the key concepts of chemical nomenclature and provide a general method for naming organic molecules, but be warned—there are plenty of exceptions that you will come across!

Naming alkanes

In order to name alkanes using IUPAC nomenclature you must:

1. Identify the parent hydrocarbon chain. This will be the longest continuous carbon chain. If there are no branches on the chain, this molecule will simply be named according to the number of carbon atoms in the chain (methyl, ethyl, etc.), followed by –ane. This is the **root name**. In the case of cyclic alkanes, cyclo- is included before the root name, e.g. cyclopropane, cyclopentane, etc.

2. If there is branching present, number the carbon atoms in the chain, ensuring the first alkyl substituent (branch) in the chain has as low a number as possible. This is the **first point of difference** rule.

3. Number and name the alkyl substituents on the chain. This will become a prefix on the parent hydrocarbon's name. If there are multiple substituents to be added to the beginning of a name, then this is done in alphabetical order.

Worked example 1.2A

Name the following alkane according to IUPAC nomenclature.

Solution

First we have to identify and number the parent hydrocarbon chain. This is not always easy to do, and can be hidden within a complex framework—so be careful! We must also ensure that the first branch has the lowest possible number on the parent hydrocarbon chain. Numbering along the chain gives us two possible arrangements, one of which has the first alkyl substituent with a lower number, so is the correct numbering when naming the compound.

correct   incorrect

This allows us to see that the parent hydrocarbon chain is eight carbon atoms long, so the name of the molecule will end –**octane**. Now we must deal with the alkyl substituents, or 'branches'. The substituent at carbon 2 (or C2) is —CH$_3$, giving a 2-methyl- prefix, while C5 has a —C$_2$H$_5$, or a 5-ethyl- prefix. We can now add these substituents in **alphabetical order** to the beginning of the parent hydrocarbon name to give **5-ethyl-2-methyloctane**.

Naming compounds with one functional group

When naming organic molecules with a single functional group we must modify our approach slightly. Now, when naming an organic molecule we must:

1. Identify the functional group present. This will give the molecule either a prefix or suffix, and in the latter case, will replace the –ane on the end of the root name. Where possible, it is typical to use the suffix, rather than prefix. Some common examples of functional group nomenclature are shown below:

Class:	alcohols	aldehydes	ketones	carboxylic acids	alkyl halides
Structure:	R—OH	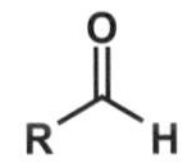 R-CHO	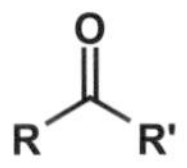 R-CO-R'	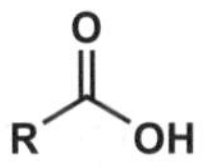 R-COOH	 R—X
Suffix:	-ol	-al	-one	-oic acid	—
Prefix:	hydroxy-	oxo-*	oxo-*	—	X: F (fluoro-) Cl (chloro-) Br (bromo-) I (iodo-)

*rarely encountered

2. Identify the longest hydrocarbon chain **to which the functional group is attached**. We now know the root name, and the suffix/prefix to be added to it. For example:

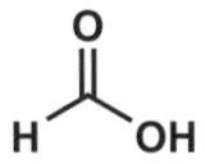

methanoic acid
or
'formic acid'

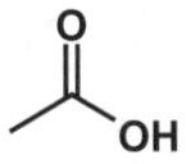

ethanoic acid
or
'acetic acid'

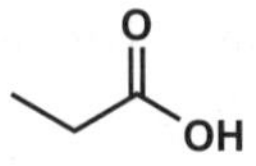

propanoic acid
or
'propionoic acid'

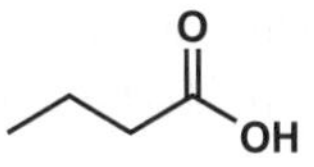

butanoic acid
or
'butyric acid'

➔ To make matters more confusing, this is not the case for alkyl halides, alkenes or alkynes, for which the parent hydrocarbon chain is the longest continuous carbon chain, as for the naming of alkanes. For instance:

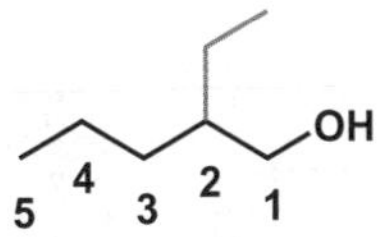

2-ethylpentan-1-ol

NOT

3-(hydroxymethyl)hexane

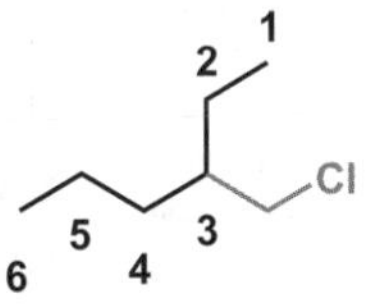

3-(chloromethyl)hexane

NOT

1-chloro-2-ethylpentane

In 2-ethylpentan-1-ol, the parent hydrocarbon chain is the longest continuous carbon chain which the functional group (—OH) is attached to. However, in 3-(chloromethyl)hexane, as it contains a halogen functional group the parent hydrocarbon chain is the longest continuous chain anywhere in the molecule. The halogen-containing chloromethyl substituent is added as a prefix onto the name. This should not crop-up too often, but is important to be aware of.

3. Number the parent hydrocarbon chain so that the functional group present has the lowest possible number. Any alkyl substituents on the molecule may then be numbered according to this functional group. For instance:

4-methylhexan-3-one

NOT

3-methylhexan-4-one

2-chloro-4-methylpentane

NOT

4-chloro-2-methylpentane

Worked example 1.2B

Use the information provided in the IUPAC name to draw out 4-ethyl-5-methylheptan-2-one.

Solution

In order to draw out this molecule, we have to extract structural information from the IUPAC name in a methodical manner. We can work backwards throughout the name to get all the information we need.

1. The suffix 2-one tells us that this is a ketone at C2.
2. The root name heptan- tells us that the parent hydrocarbon chain is seven carbon atoms long.
3. The prefix 4-ethyl-5-methyl tells us that there are ethyl and methyl substituents at the C4 and C5 positions, respectively.

Using these three pieces of information we can now start to draw out the target molecule. A good approach to this would be to draw out the parent hydrocarbon chain, number it, then place the functional group and alkyl substituents in the positions identified. Firstly, we can draw the seven-carbon chain, then number it from left to right. At the C2 position, we can now add our ketone functional group. Now we can add the ethyl group to position C4, and the methyl group to position C5. This gives us the structure 4-ethyl-5-methylheptan-2-one.

4-ethyl-5-methyl**heptan**-2-one
(7)

draw carbon chain

number carbon atoms

4-ethyl-5-methyl**heptan**-2-one

add functional group and substituents

Naming compounds with more than one functional group

If a molecule has more than one functional group, there is an important decision to be made—which functional group dictates the suffix of the IUPAC name? In order to decide this, IUPAC have prioritized certain functional groups over others, with higher priority functional groups dictating both the suffix (where appropriate) and the parent hydrocarbon chain. IUPAC call this the 'seniority' of the group. The seniority of functional groups is laid out in Appendix 3.

Worked example 1.2C

Isoleucine is an essential amino acid, meaning that humans cannot produce it themselves and must acquire it in their food. It has the IUPAC name 2-amino-3-methylpentanoic acid. Using this information, draw isoleucine.

Solution

We were told in the question that isoleucine has the alternative IUPAC name of 2-amino-3-methylpentanoic acid, which contains all the information we need to draw the structure out. We can break down the name into three parts as in Worked example 1.2B:

1. The suffix, –oic acid, tells us that this is a carboxylic acid. As carboxylic acids can only lie on terminal carbon atoms, this is by definition the C1 position.
2. The root name is pentan-, so the parent hydrocarbon chain is five atoms long.
3. The prefixes tell us the character and position of the substituents. 2-amino tells us that there is an amine group at the C2 position, while 3-methyl tells us that there is a methyl group at the C3 position.

Using this information, we can draw out the parent hydrocarbon chain, number the carbon atoms, then add the functional groups and substituents in the correct positions. The process is shown below:

➡ The prioritization of functional groups is important in deciding the suffix of the molecule name. Without prioritization there would be many different names that could be given to the same molecule. For instance, in the molecule below, IUPAC dictates that alcohols are given priority over amine groups, therefore the correct IUPAC name is 3-aminopropan-1-ol.

3-aminopropan-1-ol

or

3-hydroxypropanyl amine

2-amino-3-methylpentanoic acid

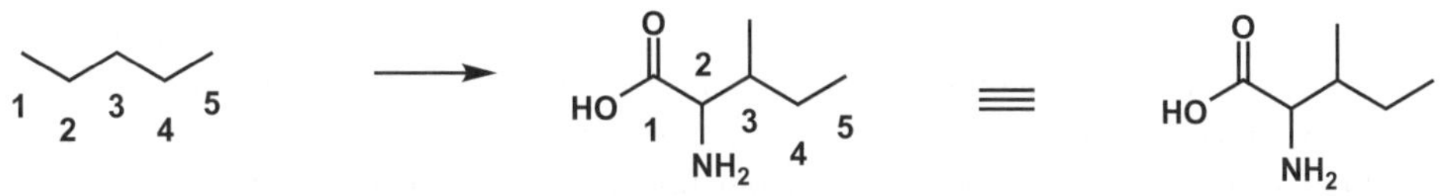

draw and number parent hydrocarbon chain → add functional groups and substituents ≡ isoleucine

---

❓ Question 1.3

Determine the IUPAC name for the following molecules.

(a) (b) 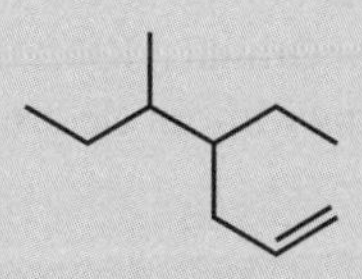 (c)

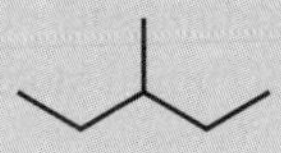

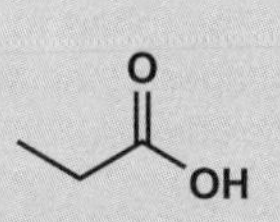

(d) 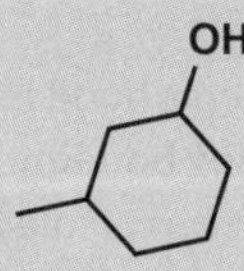 (e) 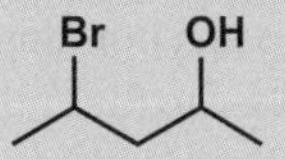 (f)

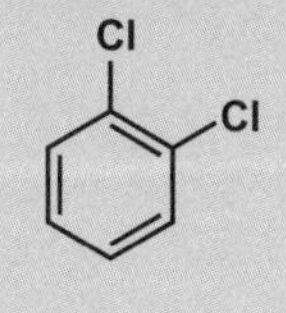

▶ **Hint** Use the nomenclature tables in Appendix 3 to help you if any functional groups are unfamiliar to you, and be careful to identify priority when more than one functional group is present.

> **❓ Question 1.4**
>
> Draw the structures of the following molecules from the information provided in their IUPAC name.
>
> (a) 4-ethyl-6,6-dimethyloctane.
>
> (b) 1,1-dimethylcyclopropane.
>
> (c) 1,3-dichloropentane.
>
> (d) 2-propyn-1-ol.
>
> (e) 1-methyl-4-(1-methylethyl)benzene (trivial name: 'cymene').
>
> (f) 3,7-dimethylocta-1,6-dien-3-ol.
>
> ▶ **Hint** When numbering -ene and -yne functional groups, the numbering is conducted from the first carbon atom of the functional group. For instance, in pent-1-ene, the double bond is between C1 and C2, and in pent-2-ene the double bond is between C2 and C3.
>
>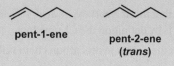
> pent-1-ene   pent-2-ene
>   (*trans*)

⮕ 3,7-dimethylocta-1,6-dien-3-ol, or 'linalool', is a popular perfume to add to shampoos and bubble-bath. It smells floral and spicy.

⮕ When we talk about orbitals being 'in-phase' or 'out-of-phase', we mean that the wave-like part of the orbital is either constructively overlapping or destructively overlapping, respectively. For more information on this, see Chapter 4 in Burrows et al. (2013).

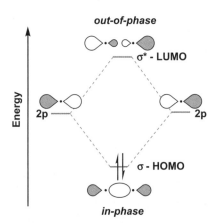

Figure 1.1 MO diagram showing the overlap of two 2p orbitals to form a σ-bond. The shapes of the orbitals involved are included.

1.3 Orbital overlap and bonding

Atoms consist of a nucleus, containing protons and neutrons, surrounded by electrons. You should be aware at this point that these electrons are contained in atomic orbitals (AOs), designated s, p, d, and f, that can sit in different shells. In organic chemistry, we mainly concern ourselves with s and p orbitals, which have characteristic sphere and dumb-bell shapes, respectively. In order to form a molecule, atoms must allow their AOs to overlap and form molecular orbitals, that bond the atoms to each other. The combination of two AOs results in the production of two molecular orbitals, one that is lower in energy than the constituent AOs (bonding), and one that is higher in energy (anti-bonding). Bonding orbitals are formed when AOs overlap in-phase, while anti-bonding orbitals are formed when they combine out of phase. The behaviour of electrons in these orbitals is described by molecular orbital (MO) theory, which we will touch upon here. The overlap of an s orbital with either an s or p orbital leads to the formation of a σ (sigma) bonding and a σ* anti-bonding MO. If a p orbital overlaps with another p orbital, two types of bonds can be formed. 'Head on' overlap of the p orbitals leads to the formation of a σ bonding and a σ* anti-bonding MO, whereas a 'side on' overlap leads to the formation of a π (pi) bonding and a π* anti-bonding MO. These bonds are weaker because the orbitals do not overlap so well. An example MO diagram is shown in Figure 1.1, which shows the 'head-on' overlap of two 2p to form a σ-bonding orbital and a σ*-anti-bonding orbital. A pair of electrons sits in the σ-bonding orbital, so a σ-bond is formed. In the MO diagram we can see the Highest Occupied Molecular Orbital (HOMO), which is the highest energy orbital with electrons in. The lowest energy orbital without electrons in is called the Lowest Unoccupied Molecular Orbital (LUMO).

Bond order

If either a σ or π bonding MO contains a pair of electrons, then a bond is formed. If a pair of electrons is added to the anti-bonding MO, then the corresponding bonding MO breaks. The number of bonds shared between two atoms, or the 'bond order', can be calculated using equation 1.1:

$$\text{Bond order} = \frac{\text{bonding electrons} - \text{anti-bonding electrons}}{2} \quad (1.1)$$

where 'bonding electrons', and 'anti-bonding electrons' refers to the number of electrons in bonding and anti-bonding orbitals, respectively. This value must always be zero or a positive integer, so if your answer is negative, something has gone wrong! We can see that if we would add a pair of electrons to the σ bonding orbital in Figure 1.1, we would have to occupy the LUMO—the σ* anti-bonding orbital (Figure 1.2). This would 'break' the σ-bond, and reduce the bond order to zero. This will be very important when mechanisms are covered later in the workbook.

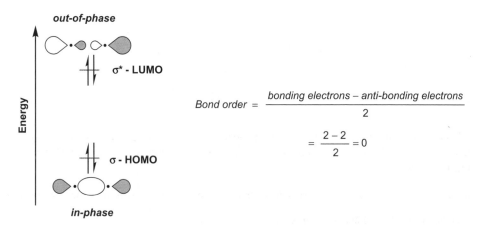

Figure 1.2 Adding a pair of electrons into the σ* anti-bonding MO reduces the bond order to zero, breaking the σ bond.

Worked example 1.3A

Show on an MO diagram what molecular orbitals are formed (if any) from the constituent AOs of H_2. Calculate the bond order.

Solution

This is the simplest possible scenario when drawing these MO diagrams, but care still needs to be taken. First, we must identify what orbital the valence electron of hydrogen is in—namely the 1s orbital. We can then draw the MO diagram, showing the energy level of the 1s orbital for each hydrogen atom—this will be the same in each atom. We can then add an electron into each of these orbitals, completing the electronic arrangement for atomic H.

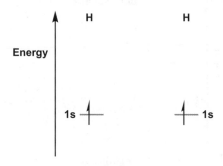

→ There are three important rules to remember when adding electrons to MO diagrams:

- The Aufbau principle, which means that you start filling the lowest energy orbitals first.
- The Pauli exclusion principle, which states that each orbital may only contain two electrons, of opposite spin.
- Hund's Rule, which states that when there are orbitals which are equal in energy (degenerate), electrons are added one at a time to each orbital, before pairing.

We can now draw the molecular orbitals of H_2 in the centre of the diagram. The combination of the constituent 1s orbitals will result in **two** new MOs—a σ bonding orbital, and a σ* anti-bonding orbital. The bonding orbital is lower in energy than the anti-bonding orbital, and the 1s orbitals. To complete the MO diagram, fill the orbitals from bottom to top with the electrons from the constituent orbitals. This will demonstrate that a σ-bond is being filled.

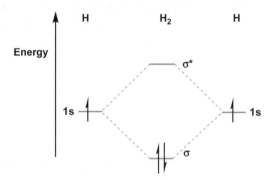

Finally, to calculate the bond order, simply use the equation given previously. In this case, we can see from the diagram that there are two electrons in bonding orbitals, and no electrons in anti-bonding orbitals. So:

bond order = (number of electrons in bonding orbitals—number of electrons in anti-bonding orbitals)/2

bond order = (2−0)/2 = 1 i.e. a new single bond.

Worked example 1.3B

Draw the MOs arising from the overlap of a 2s orbital with another 2s orbital.

Solution

We know from the introductory paragraph that when s orbitals overlap with s orbitals they form σ-bonds. This is also the case when a 2s orbital overlaps with another 2s orbital while in-phase. However, we must be careful to remember that when two new AOs combine they produce two MOs as a result. The other orbital formed is a σ* anti-bonding orbital, resulting from the out-of-phase combination of the two 2s orbitals. This can be depicted on an MO diagram, which nicely shows the energy levels of the bonding and anti-bonding orbitals:

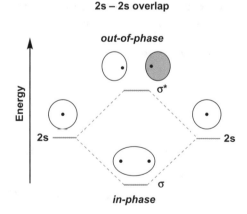

Note that in this case we have drawn the 2s orbitals as being equivalent in energy—this is not always the case, especially if the two atoms are different. However, even if the 2s orbitals are not equal in energy, the bonding orbital formed will be lower in energy than the constituent AOs. It is also the case that the anti-bonding orbital formed will be higher in energy than both of the two constituent AOs.

In future work, unless instructed, we will simply depict the resulting MOs pictorially, without an energy level diagram. This will help simplify diagrams as the examples used become more complex in later chapters. We can show the MOs for in-phase or out-of-phase 2s orbital overlap more clearly in this way:

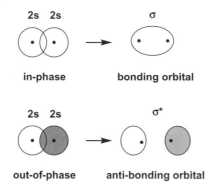

> You may have wondered why some orbitals are shaded, while others are not. This shading designates a difference in the sign of the wavefunction of the orbitals—one orbital is plus, the other is minus. A deep understanding of wavefunctions is not yet necessary, but you will need to know that for orbitals to be 'in-phase', their shading, i.e. the sign of their wavefunction, must be the same.

 Question 1.5

Show on an MO diagram what molecular orbitals are formed (if any) from the overlap of constituent AOs of He_2, and calculate the bond order.

 Question 1.6

Draw the MOs arising from:

(a) The head-on overlap of a 2p orbital with another 2p orbital.
(b) The side-on overlap of a 2p orbital with another 2p orbital.

> The MO theory section of this workbook is restricted to the key concepts which are necessary for use in organic chemistry. More information and questions on MO theory can be found in Clayden et al. (2012) and the *Workbooks in Chemistry: Physical Chemistry* book in this series

1.4 Orbital hybridization

The AOs, s, p, d, and f, are useful for describing the ground state of atoms, but often cannot explain the geometry of bonds in molecules, which arise from the overlap of these individual AOs. In order to explain observed deviations in the bond angles expected from AOs, Linus Pauling introduced the idea of 'hybrid' orbitals, which are the result of the mixing, or 'hybridization', of AOs. In this hybridization process, AOs combine to form an equal number of hybrid orbitals, which are equivalent in energy, or 'degenerate', shown in Figure 1.3. These hybrid orbitals have some of the character of each of the AOs that hybridized to form them. Although all elements are theoretically able to hybridize, in organic chemistry we mainly concern ourselves with the hybridization of carbon, especially for introductory courses. The valence electrons in carbon sit in the 2s and 2p orbitals, and these can hybridize to form sp, sp^2, or sp^3 hybrid orbitals, depending on the bonding of the atom. The hybridization of one s and three p orbitals gives rise to sp^3 orbitals; sp^2 orbitals arise from one s and two p orbitals, and sp orbitals arise from one s and one p orbital. These hybrid orbitals retain some of the character of their constituent orbitals. An overview of these hybridization processes is provided in Figure 1.3. These hybrid orbitals can overlap 'head-on' with s, p, and other hybrid orbitals to form σ bonding orbitals.

> Further information on the reasons for orbital hybridization can be found in Clayden et al. (2012).

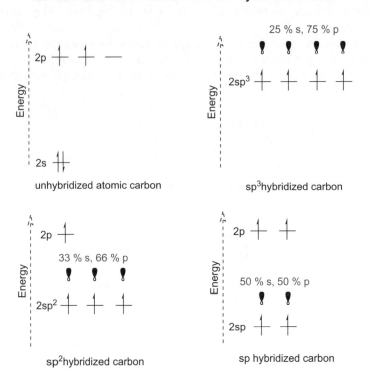

→ Note that electrons will spread evenly throughout orbitals of equivalent energy, according to Hund's rule.

Figure 1.3 MO diagrams showing the energy levels and shapes of carbon in its various stages of hybridization. The percentages of s and p character are inlaid.

It is straightforward to identify the hybridization of carbon atoms in a molecule, as long as they possess a full valence shell and are uncharged, which is most commonly the case.

→ Note that in these images we have shown the AOs of the constituent atoms. Looking at the molecule as a whole, these AOs will overlap and form MOs (bonds), it is just useful to see them in this state to understand the resulting geometry of the molecule.

- If the carbon atom contains four σ (single) bonds, and no π (double) bonds then it is sp³ hybridized, and will be tetrahedral, e.g. the carbon atom in methane.

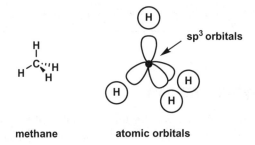

methane   atomic orbitals

→ Note that in this case, the sp² orbitals are perpendicular to the p-orbital, i.e. the p-orbital points vertical, and the sp² orbitals 90° to it

- If the carbon atom contains three σ-bonds, and one π-bond then it is sp² hybridized, and will be trigonal planar, e.g. the carbon atoms in ethene.

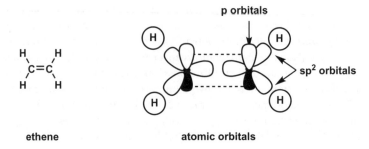

ethene   atomic orbitals

- If the carbon atom contains two σ-bonds, and two π-bonds then it is sp hybridized, and will be linear, e.g. the carbon atoms in ethyne.

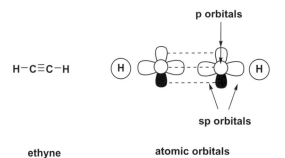

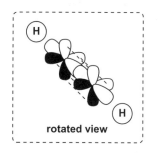

> Note that a molecule is able to rotate around a single bond, however, when you have π-bonds, this rotation cannot occur. This is very important in stereochemistry, covered in Chapter 2.

Hybridization can also occur in other elements, such as oxygen and nitrogen. Similar to carbon, whether the atom contains a double bond or triple bond is a good indicator that it is likely to be sp² or sp hybridized. Atoms which are not carbon are more likely to possess vacant orbitals or lone pairs of electrons, and in this case, we have to decide which orbital any lone pairs of non-bonding electrons will reside in. If the lone pair is not conjugated, then it will likely reside in a hybridized orbital. For conjugated lone pairs of electrons, they will sit in p-orbitals in order to provide better overlap with any conjugated π electrons. For instance, the nitrogen atom in an amine is sp³ hybridized, whereas the nitrogen atom in an amide group is sp² hybridized.

Worked example 1.4A

Label the carbon atoms in the following molecule as sp, sp², or sp³ hybrids.

(S)-2-Amino-4-pentynoic acid

Solution

(S)-2-Amino-4-pentynoic acid, also known as L-propargylglycine, contains five carbon atoms, numbered 1–5 from the carboxylic acid.

Looking at the structure of the molecule, we are able to divide these carbon atoms into three groups: those with only single bonds; those containing a double bond; and those containing a triple bond. Carbons 2 and 3 contain only σ-bonds, thus they are **sp³ hybridized**. Carbon 1 contains a double bond, consisting of one π-bond and one σ-bond, as well as two additional σ-bonds, thus it is **sp² hybridized**. Finally, carbons 4 and 5 are triple bonded to one-another, so both have two σ-bonds and two π-bonds, and are **sp hybridized**.

Worked example 1.4B

What is the hybridization state of the carbon atoms in propadiene? Draw the AOs, showing the overlap that will occur to form MOs (bonds).

$$H_2C=C=CH_2$$
propadiene

Solution

First, we will identify the hybridization state of the carbon atoms present. The outer two carbon atoms have three σ-bonds and one π-bond each. Thus, they are **sp² hybridized**. The central carbon may look a little unusual, but looking at the molecular orbitals present will lead you to the correct hybrid state. It is in possession of two σ-bonds and two π-bonds, and is therefore **sp hybridized**.

Next, we are asked to draw the AOs involved in the bonding of propadiene. Using the knowledge of the hybridization states from the first part, we now simply need to draw the characteristic shapes of sp² and sp hybridized carbons in the order that they are in pentadiene.

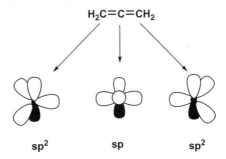

We must think about the orientation of these AOs a bit more carefully. In order to have good enough orbital overlap to form a π-bond, p-orbitals must be aligned with one another. To achieve this, we must rotate one of the outer sp² hybridized carbon atoms so that the p-orbital is in-line with the p-orbital of the central sp hybridized carbon atom, i.e. facing the viewer.

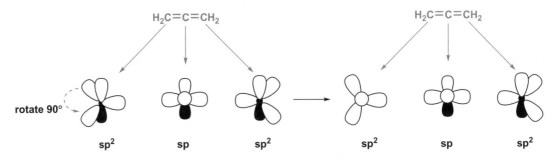

Finally, we can move the orbitals closer to show the sites of overlap, and include the two orbitals of the hydrogen atoms attached to the outer carbon atoms. Here, dashed lines are used to show how the p-orbitals would overlap to form a π-bond.

$$H_2C=C=CH_2$$

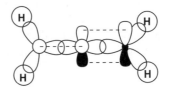

1.5 DOUBLE BOND EQUIVALENTS

Question 1.7

Label all carbon atoms in the following molecules as sp, sp^2, or sp^3 hybridized.

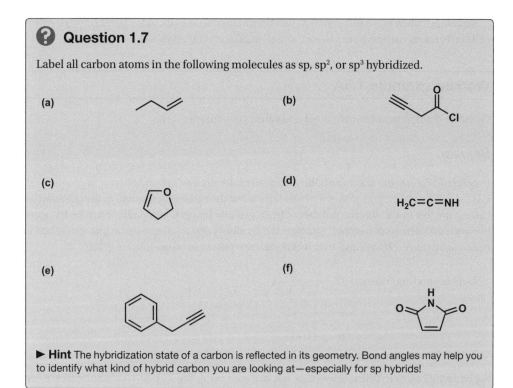

▶ **Hint** The hybridization state of a carbon is reflected in its geometry. Bond angles may help you to identify what kind of hybrid carbon you are looking at—especially for sp hybrids!

Question 1.8

Label all non-hydrogen or carbon atoms in the following molecules as sp, sp^2, or sp^3 hybridized.

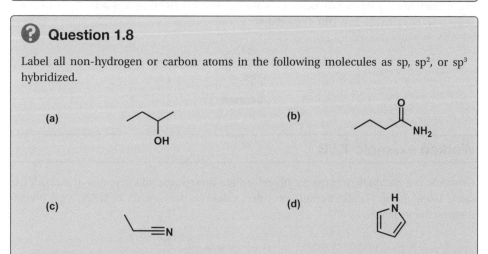

▶ **Hint** Nitrogen and oxygen atoms that don't possess any π-bonding can still be sp^2 hybridized if they are adjacent to a π-bond (conjugated). This does not happen for saturated carbon atoms as they do not possess a lone pair of electrons.

1.5 Double bond equivalents

When we are trying to determine the structure of a molecule from its molecular formula, it is useful to be able to know the number of π-bonds and cyclic structures in the molecule. This can be achieved by calculation of the double bond equivalents, also called the 'degree of unsaturation' in some textbooks. The formula for calculating this is shown in equation 1.2:

$$\text{Double bond equivalents (DBEs)} = C - \frac{H}{2} + \frac{N}{2} + 1 \tag{1.2}$$

Where C = number of carbon atoms, H = number of hydrogen atoms **and halogen** atoms, and N = number of nitrogen atoms.

→ The calculation of DBEs using equation 1.2 will be incorrect for molecular formulae containing atoms in higher valence states, such as N(V), P(V), and P(VII), so don't be surprised if this formula does not work for every molecule.

Remember when using this equation that the number of DBEs includes all π-bonds, so covers carbonyl groups, carboxylic acids, etc., as well as alkenes and alkynes.

Worked example 1.5A

Calculate the number of double bond equivalents in benzene, C_6H_6.

Solution

In order to calculate the number of DBEs, we need to identify the C, H, and N terms in equation 1.2. C is the number of carbon atoms, and looking back at the molecular formula in the question we can equate this to six. H is the number of hydrogen and halogen atoms. There are no halogens present, and six hydrogen atoms. Therefore the H value is also six. There are no nitrogen atoms in benzene, so the N term is equal to zero. We can now put these values into equation 1.2:

$$\text{Double bond equivalents} = C - \frac{H}{2} + \frac{N}{2} + 1$$
$$= 6 - \frac{6}{2} + \frac{0}{2} + 1$$
$$= 6 - 3 + 0 + 1$$
$$= \mathbf{4}$$

So, benzene has four DBEs. If we draw out benzene using the Kekulé-type structure, we can count that it has three π-bonds. Benzene is also cyclic, which is equal to one DBE. These observations account for the four DBEs calculated.

→ The molecular formula of benzene was known long before its structure was determined. In addition to Kekulé's structure, there were conflicting structures proposed by chemists including Adolf Claus and James Dewar. Kekulé's structure is now used to depict benzene, but be aware that due to electron delocalization (aromaticity), it does not tell the whole story. This will be covered later in the chapter.

Kekulé

Dewar

Claus

benzene

Worked example 1.5B

Capsaicin is a naturally occurring mucosal irritant present in chilli peppers—it makes them spicy! Using the molecular formula provided, calculate the number of DBEs, then identify them on the structure.

Capsaicin ($C_{18}H_{27}NO_3$)

Solution

We can take the C, H, and N values from the molecular formula given above, equal to 18, 27, and 1, respectively. If we put these into equation 1.2 we get:

$$\text{DBEs} = C - \frac{H}{2} + \frac{N}{2} + 1$$
$$= 18 - \frac{27}{2} + \frac{1}{2} + 1$$
$$= 18 - 13.5 + 0.5 + 1$$
$$= \mathbf{6}$$

1.5 DOUBLE BOND EQUIVALENTS

So capsaicin has six DBEs. Now we need to identify them on the molecule. The number of DBEs is equal to the number of π-bonds plus the number of cyclic structures. Looking at capsaicin we can see five π-bonds—four alkene groups and one carbonyl group. There is also one six-membered ring on the right hand side of the molecule. This accounts for the sixth DBE.

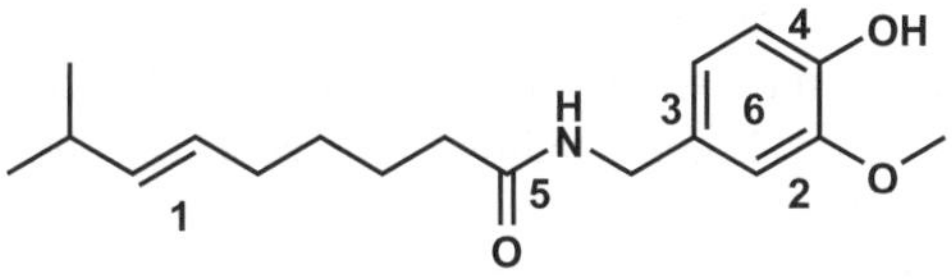

Question 1.9

Calculate the number of double bond equivalents from the following molecular formulae.

(a) C_4H_8O
(b) $C_6H_{12}O_6$
(c) C_4H_7NO
(d) $C_6H_4Cl_2$
(e) $C_{20}H_{12}O_5$ (fluorescein, a fluorescent dye)
(f) $C_5H_{13}ClN_2O$
(g) $C_2HF_3O_2$
(h) $C_{20}H_{14}N_4$ (porphine, a metal-binding ligand present in many proteins)

▶ **Hint** Don't forget that the H term in the DBE equation counts hydrogen **and halogen atoms**.

Question 1.10

Tosyl chloride is commonly used to add a tosyl- protecting group to amines and alcohols. By simply counting, we can see that the number of DBEs is 6.

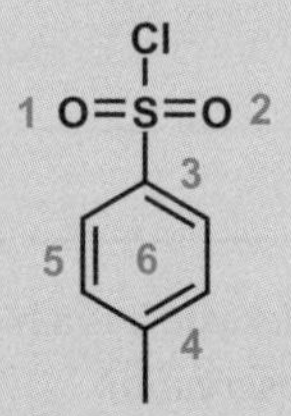

tosyl chloride
($C_7H_7ClO_2S$)

However, using equation 1.2, the number of DBEs appears to be 4:

$$\text{tosyl chloride DBEs} = 7 - \frac{8}{2} + \frac{0}{2} + 1 = 7 - 4 + 0 + 1 = 4$$

Which value is correct, and why?

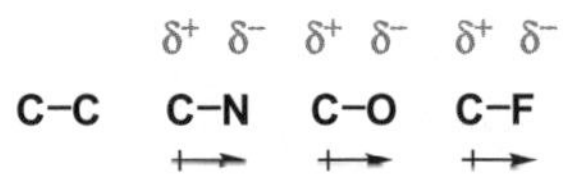

δ^+ δ^- δ^+ δ^- δ^+ δ^-
C—C   C—N   C—O   C—F
      ⟼   ⟼   ⟼

Figure 1.4 Polar bonds between carbon (χ: 2.5) and nitrogen (χ: 3.0), oxygen (χ: 3.5), and fluorine (χ: 4.0). The C—C bond is apolar due to the equal electronegativity of the carbon atoms.

➲ The arrow notation used here is the most commonly used. However, you will also see a non-crossed arrow used to indicate polarity, which points from δ^- to δ^+. This is actually what IUPAC says chemists should use, but at the time of writing, it has not been widely adopted.

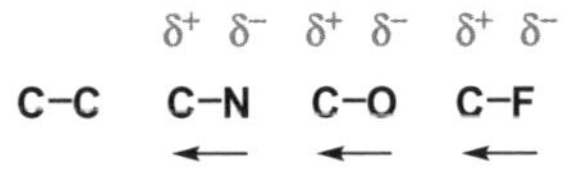

δ^+ δ^- δ^+ δ^- δ^+ δ^-
C—C   C—N   C—O   C—F
      ⟵   ⟵   ⟵

1.6 Polarity

In bonds between two different chemical elements, the electrons aren't always equally attracted to each atom. They can be drawn towards certain elements more than others, depending on that element's electronegativity (χ). More electronegative elements are able to draw electrons in bonds towards them from less electronegative elements, resulting in a slight negative charge (δ^-) at their end of the bond, and a slight positive charge (δ^+) on the other. This creates a dipole and is indicated by a crossed arrow pointing from the δ^+ end to the δ^- (Figure 1.4). The electronegativity of elements is available through electronegativity tables but generally increases in magnitude as you go across a period and decreases as you go down a group on the periodic table. Please note that those elements with electronegativity lower than carbon (χ: 2.5) are often referred to as electropositive, and 'push' electrons towards carbon.

So, bonds can have polarity, but we must be aware that molecules also have an overall molecular polarity, or **dipole moment**, which depends on the polarity of the bonds within the molecule and their arrangement in space. For instance, the C—Cl bond is polar, which results in dichloromethane having overall polarity. However, due to the arrangement of C—Cl bonds in tetrachloromethane, it has no molecular polarity (Figure 1.5). A dipole moment has both magnitude and direction, determined by the arrangement of polar bonds within the molecule. Lone pairs of electrons residing on atoms also contribute to polarity.

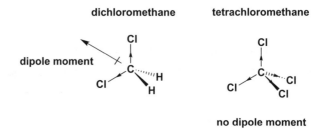

Figure 1.5 Polar bonds can cause a molecule to have an overall dipole moment, as in dichoromethane. However, if these bonds oppose each other, as in tetrachloromethane, they can cancel each other out, leaving the molecule with no dipole moment.

Worked example 1.6A

Despite having similar structures, formaldehyde has a high dipole moment, while carbon dioxide has no dipole moment. Explain this observation.

formaldehyde
(polar)

carbon dioxide
(nonpolar)

Solution

In order to understand the overall polarity of the molecules, we must look at the polarity of the bonds present. Formaldehyde possesses C—H and C=O bonds. Carbon and hydrogen have similar electronegativities (χ: 2.5 and 2.2, respectively), so their bond is not considered to be polar. The electronegativity of oxygen is greater than carbon (χ: 3.5 and 2.5, respectively), so the carbonyl group is polar, with a δ⁻ charge on the oxygen. The polarity of this bond gives formaldehyde an overall dipole moment. Carbon dioxide possesses two polar carbonyl groups. However, these directly oppose each other symmetrically. This means that overall, carbon dioxide is not polar.

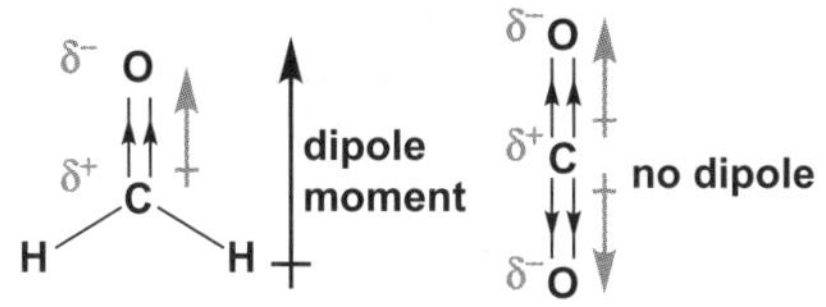

> Although we have said here that the C—H bond is not polar, there is still a slight difference in electronegativity between carbon and hydrogen. This slight dipole on the C—H bond may not be noticeable in some cases, but in alkyl substituents the δ⁻ charge can build up on carbon atoms through multiple C—H bonds, making the charge much more significant. This means that overall, alkyl groups are electron donating by induction.

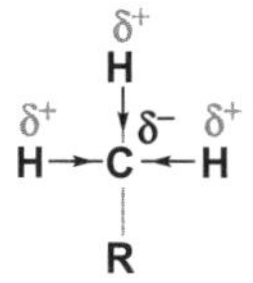

The inductive effect

Polarization across a bond can have a knock-on effect along a chain of atoms, 'pulling' or 'pushing' electron density along σ-bonds. This effect diminishes along a chain of atoms the further you get from the polar bond. This is called the inductive effect (Figure 1.6). Atoms or groups which 'pull' electron density along σ-bonds are called 'electron withdrawing', and designated '−I'. Atoms which 'push' electron density along σ-bonds are called 'electron donating' or 'electron releasing', and designated '+I'. This inductive effect can serve to modulate the pK_a of organic acids, and stabilize charge. This is especially important when dealing with carbocations, where the presence of +I groups can greatly improve stability.

> For more information on pK_a, see Chapter 3, section 3.4.

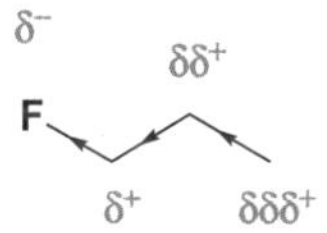

Figure 1.6 An electronegative fluorine atom is able to draw electron density through σ- bonds. This is known as the inductive effect.

> Electron withdrawal and donation can also occur through a π system, as described by the mesomeric effect. See section 1.8 for details.

Worked example 1.6B

Acetic acid has a pK_a of 4.75, while 2-chloroacetic acid is more acidic, with a pK_a of 2.85. Explain this difference in acidity.

Would you expect 2,2-dichloroacetic acid to have a higher or lower pK_a?

acetic acid
pK_a: 4.75

2-chloroacetic acid
pK_a: 2.85

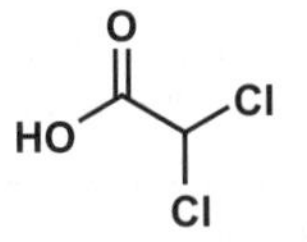

2,2-dichloroacetic acid
pK_a: ?

Solution

In order to answer this question we need to know what factors affect the pK_a of organic acids. The more stable the conjugate base (the molecule after deprotonation), the more readily it will dissociate from its proton, and the lower the pK_a will be. For uncharged molecules like those above, the stability of the conjugate base is determined by its ability to stabilize the resulting charge. So, when we compare the conjugate base of acetic acid to that of 2-chloroacetic acid we need to look at what factors could affect the stability of the negative charge on the oxygen atom. 2-chloroacetic acid has a chlorine atom at the α-position. Chlorine is more electronegative than carbon, so will draw electron density towards itself through σ-bonds. This effect can continue through to the oxygen atom, reducing the negative charge residing on that atom through induction. This stabilization is enough to cause this increase in acidity, and associated reduction in pK_a. It should be noted that the methyl group on acetic acid actually serves as a +I substituent, so destabilizes the charge on the conjugate base of acetic acid.

→ It is important to note that charge can also be stabilized by resonance, which is covered in section 1.8.

acid

acetic acid
pK_a: 4.75

2-chloroacetic acid
pK_a: 2.85

conjugate base

+I effect
less stable

−I effect
more stable

Now, having justified increased acidity due to inductive effects in the first part of this question we can try to make a prediction about the acidity of 2,2-dichloroacetic acid. The inductive effect is increased in the presence of additional −I groups. This means that an additional chlorine atom will serve to further draw electron density away from the charged oxygen atom, stabilizing the conjugate base of 2,2-dichloroacetic acid, so it is likely to be more acidic, and have a lower pK_a. This **is** the case, as the literature reports that 2,2-dichloroacetic acid has a pK_a of 1.35.

> ### Question 1.11
>
> Are the following molecules polar or nonpolar?
>
> (a) CF_4
>
> (b) HCN
>
> (c) BCl_3
>
> (d) SO_2
>
> (e) $(CH_3)_2SO$
>
> ▶ **Hint** Revisit VSEPR theory for help deciding the geometry of these molecules. Then decide if there is any symmetry that might affect the polarity of the molecules.

> ### Question 1.12
>
> Which of the following charged molecules would you expect to be more stable?
>
> (a)
>
> A   or   B

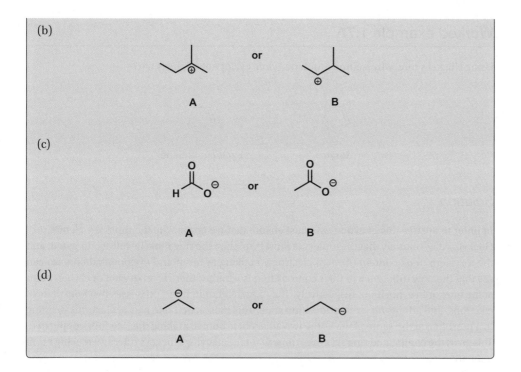

1.7 Aromaticity

We have seen in section 1.3 that p-orbitals can overlap side-on to form π-bonds. If two or more of these π-bonds are adjacent to each other in an organic molecule, and their p-orbitals are in the same plane, they are said to be conjugated. This is also the case for electrons lying in adjacent p-orbitals, such as a lone pair in an sp^2 hybridized oxygen atom. The electrons within conjugated systems can delocalize over this extended π-system, providing additional stability to conjugated systems.

Molecules that contain a planar, conjugated ring of sp^2 hybridized atoms with a delocalized electron cloud in their π-system are said to be aromatic. The delocalized π-electrons in an aromatic π-system do not behave like electrons in a non-aromatic system, and will not undergo the same chemical reactions. In order for a cyclic system to be aromatic it must be planar, contain sp^2 hybridized atoms in its ring so that all p-orbitals are parallel, and will satisfy Hückel's rule. Hückel's rule states that in order to be aromatic, a planar conjugated ring must contain $(4n + 2)$ π-electrons, where n is zero, or a positive integer (Figure 1.7). It is important to note, however, that Hückel's rule is not always effective for predicting the aromaticity of molecules containing more than one ring.

➔ You will come across some compounds that are anti-aromatic. These molecules should not be confused with non-aromatic compounds. Anti-aromatic molecules possess a conjugated ring with $4n$ π-electrons, where n is a positive, non-zero, integer. This anti-aromaticity is destabilizing, and often occurs in reactive, short-lived, molecules.

Figure 1.7 Using Hückel's rule to determine aromaticity. Benzene is planar and possesses a fully conjugated ring, which contains 6 π-electrons ($n = 1$). Benzene obeys Hückel's rule, so is considered aromatic.

Worked example 1.7A

Using Hückel's rule, why is furan aromatic, while cyclopentadiene is not?

furan

cyclopentadiene

Solution

In order to answer this question, we must ensure that we appreciate the three key elements of Hückel's rule—namely that the molecule must be planar, the ring must be fully conjugated, and the π-system must contain $(4n + 2)$ electrons. Looking at furan and cyclopentadiene, we can see that the only difference in their composition is whether there is an oxygen or carbon atom at the one, or five, position, respectively. The oxygen atom in furan possesses two lone pairs of electrons, and adopts an sp^2 hybridization state. This means that one pair of electrons is sitting in a p-orbital, in the same plane as the two adjacent π-bonds, making the ring fully conjugated. This gives the conjugated ring six π-electrons in total, satisfying Hückel's rule, and making furan aromatic.

Cyclopentadiene does not possess a fully conjugated ring due to the carbon atom at the 5-position containing no lone pairs of electrons. Additionally, the π-bonds of cyclopentadiene only contain 4 electrons ($n = 0.5$, which is not an integer), so it is non-aromatic.

→ This example illustrates that the hybridization state of heteroatoms can change in order to make a compound aromatic. Oxygen atoms in, for instance, an alcohol are sp^3 hybridized, but in furan adopt an sp^2 hybridization in order to complete the π-system. This can also happen for other heteroatoms, such as nitrogen and sulfur.

 ≡
planar
ring conjugated
6 π-electrons (n=1)
aromatic

furan

planar
ring not fully conjugated
4 π-electrons
Non-aromatic

cyclopentadiene

Worked example 1.7B

Cyclooctatetraene consists of eight sp^2 hybridized carbon atoms in a ring, possessing four π-bonds. It is conjugated and cyclic, however, it is non-aromatic and non-anti-aromatic. Why is cyclooctatetraene non-aromatic? And why is it not anti-aromatic?

cyclooctatetraene

Solution

While it may look like cyclooctatetraene is aromatic on first glance, in order to assign it so we must make sure that it obeys Hückel's rule. The four conjugated π-bonds in this molecule contain eight π-electrons. Trying to put this into Hückel's rule gives us $n = 1.5$, which is not an integer. This means that cyclooctatetraene is non-aromatic.

There is, however, an additional subtlety within this question—why is cyclooctatetraene not anti-aromatic? For compounds to be anti-aromatic they must possess a planar conjugated ring, containing 4n π-electrons. Cyclooctatetraene is cyclic, conjugated, and does contain 4n π-electrons (n = 2). Cyclooctatetraene is, however, not planar, so cannot be anti-aromatic. The reason for cyclooctatetraene not adopting a planar conformation, despite its constituent carbon atoms all being sp² hybridized is that the ideal bond angle inside an octagon is 135°, whereas the ideal bond angle on an sp² hybridized atom is 120°. This leads to the molecule adopting a non-planar 'tub' conformation due to ring strain.

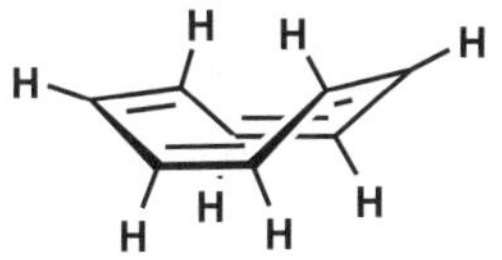

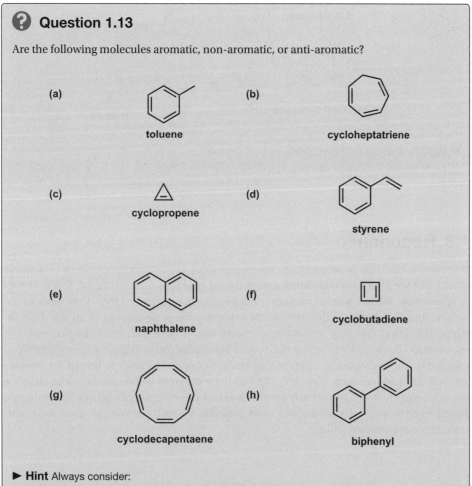

Question 1.13

Are the following molecules aromatic, non-aromatic, or anti-aromatic?

(a) toluene

(b) cycloheptatriene

(c) cyclopropene

(d) styrene

(e) naphthalene

(f) cyclobutadiene

(g) cyclodecapentaene

(h) biphenyl

▶ **Hint** Always consider:
- Does the ring only contain sp² hybridized atoms?
- Is the ring planar?
- Then, for monocyclic systems, does this ring obey Hückel's rule?

▶ **Hint** For bicyclic molecules, Hückel's rule does not necessarily apply. It may help to consider the conformation of each ring, and hybridization state of each atom in the ring to determine aromaticity for these compounds.

> **Question 1.14**
>
> Are the following heterocyclic compounds aromatic, non-aromatic, or anti-aromatic?
>
>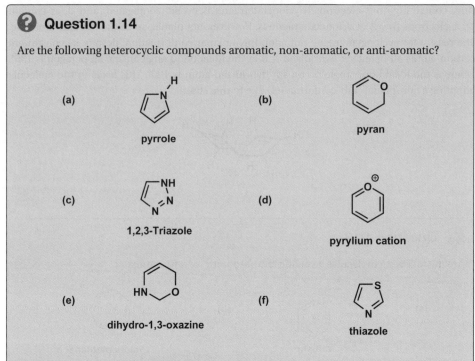
>
> ▶ **Hint** Think about the hybridization state of the heteroatoms in each of these molecules—do they contain the p-orbitals necessary to form a delocalized π-system?

1.8 Resonance

There are times, like in aromaticity, where electrons aren't simply localized to a single orbital, but are able to occupy numerous different positions on a molecule, this is known as **resonance**. We are able to visualize this by drawing different Lewis structures of the same compound, but in reality the electron distribution is an average of all the different forms. This effect can help to distribute charge over a molecule, increasing its stability. For instance, the sulfate (2−) anion has several resonance forms (Figure 1.8), which help to stabilize it. The increased stability of this anion through resonance is one of the reasons that sulfuric acid has such a low pK_a. We can draw different resonance forms by showing the motion of electrons using curly arrows, and link the different resonance forms using a double-headed arrow. Generally, the more resonance structures you can draw for a molecule, the more stable it will be.

Figure 1.8 Resonance structures of the sulfate anion.

1.8 RESONANCE

→ Sometimes you will see the resonance forms of a molecule combined into a 'resonance hybrid' structure. This shows the resonating electrons averaged as a dotted line across the molecule. For instance, in the formate anion:

resonance structures   resonance hybrid

Not all resonance structures are of equal stability, and those that are more stable have a greater contribution to the resonance hybrid. The first thought when assigning stability should be whether the molecule satisfies the octet rule, and then stabilizing effects such as electronegativity can be considered.

The mesomeric effect

The resonance, or delocalization, of electrons through π-bonds or p-orbitals in a molecule can serve to donate or withdraw electron density, which can, in turn, affect the stability or reactivity of a molecule. This is called the mesomeric (or resonance) effect, and groups which donate electron density through the mesomeric effect are designated +M, and those which withdraw electron density are –M (Figure 1.9). Remember that the actual distribution of electrons is an average of these resonance forms, and there is permanent polarization via this process. So, for neutral molecules, while you will usually draw the uncharged species, the charged resonance structure does make a contribution to the resonance hybrid, and therefore the polarity.

Figure 1.9 Some examples of –M and +M functional groups.

Worked example 1.8A

Draw all the possible resonance structures of 3-methylphenol, also known as *m*-cresol. What does this imply for the hybridization of oxygen?

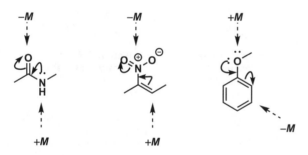

m-cresol

> Aromatic rings, such as that found in phenol, are particularly good at stabilizing charge through resonance. This is one reason why phenol is so acidic—the phenolate anion is stabilized through resonance, which weakens the O—H bond.

Solution

We can use some steps for drawing out resonance structures in a methodical manner. Firstly, we will show the movement of a pair of electrons using a double-headed curly arrow. Then, we must consider if there needs to be any further movement of electrons (usually from π-bonds) so that no atoms possess more than eight electrons, in accordance with the octet rule. Finally, we need to assign any charge on the molecule. We can draw out the first resonance structure for phenol by moving a pair of electrons from the oxygen atom to form a π-bond with the adjacent carbon atom. This would leave that carbon pentavalent (which we must never do!), with ten electrons in its outer shell. So, we must move a pair of electrons from that carbon atom's other π-bond onto the neighbouring C2 position. This process would leave oxygen with a positive charge, having been made to share an electron with the carbon at position 1, and the carbon at position 2 with a negative charge, having 'gained' an electron from the previously-shared π-bond. We can continue this process around the aromatic ring to give 2 additional resonance structures. As a lone pair on the oxygen is able to conjugate with the π system, the oxygen atom must be sp^2 hybridized.

Worked example 1.8B

Acetone has a pK_a of 19.2, whereas the hydrogen atom at the 3-position of acetylacetone has a pK_a of 9.0. Explain this difference in acidity.

propan-2-one
"acetone"
pKa : 19.2

pentane-2,4-dione
"acetylacetone"
pKa : 9.0

Solution

The pK_a of organic molecules is related to the stability of the conjugate base, as discussed in Worked example 1.6B. If we draw out the conjugate base of acetone and acetylacetone, we can then try to rationalize which is likely to be more stable.

acid

conjugate base

1.8 RESONANCE

Now, we know from sections 1.6 and 1.8 that the main contributors to stability are inductive effects and resonance. As in Worked example 1.6B, there is likely to be a contribution from the inductive effect, due to electron withdrawal by the oxygen atom of the carbonyl groups. However, resonance effects are typically more stabilizing than inductive effects, so we must consider these as well. We shall first draw out all possible resonance structures of acetone and acetylacetone. This gives us two resonance structures for acetone, and three for acetylacetone. This means that the conjugate base of acetylacetone is more stable than that for acetone, so acetylacetone has a lower pK_a. It should be noted, also, that resonance forms which leave a negative charge on an electronegative atom are particularly stable.

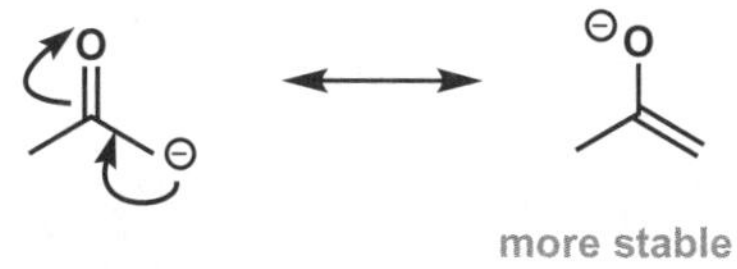

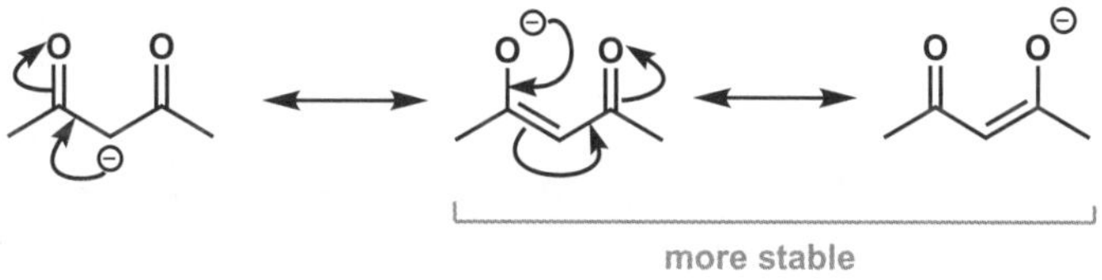

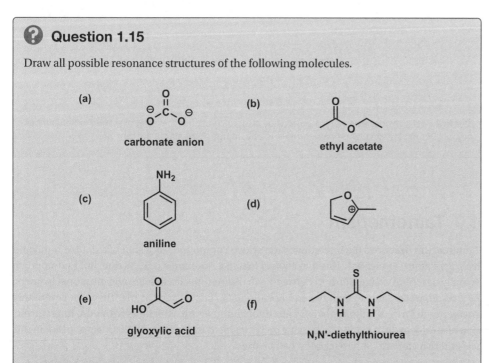

Question 1.15

Draw all possible resonance structures of the following molecules.

(a) carbonate anion

(b) ethyl acetate

(c) aniline

(d)

(e) glyoxylic acid

(f) N,N'-diethylthiourea

▶ **Hint** Be systematic, starting at a pair of non-bonded electrons moving to create a new π-bond. Then, move electrons from π-bonds to new orbitals to ensure that there are no pentavalent species created. Finally, decide if the movement of electrons has created any charged atoms. It will also help to draw these as Lewis structures, displaying the lone pairs of electrons as dots.

> **Question 1.16**
>
> In each case, identify the most stable resonance structure (the 'major contributor') for the following molecules. Justify your choices.
>
> (a) i. ↔ ii.
>
> (b) i. ↔ ii.
>
> (c) i. ↔ ii.
>
> (d) i. N=C=O ↔ ii. N≡C−O⁻
>
> ▶ **Hint** First consider whether the resonance structure satisfies the octet rule, then whether it has fewer charged species, and finally, if there is any charge, how stable the ions are.

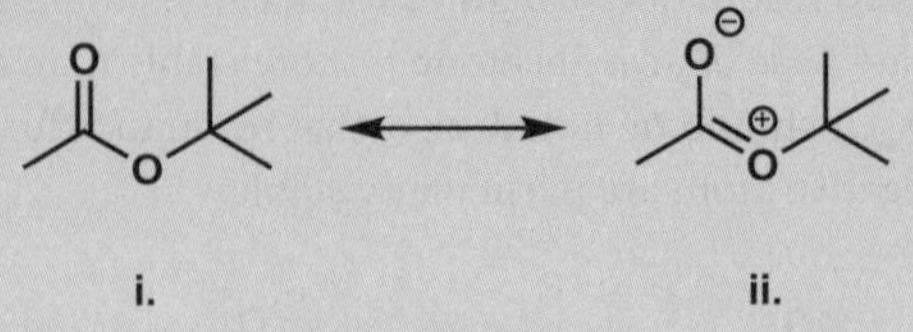

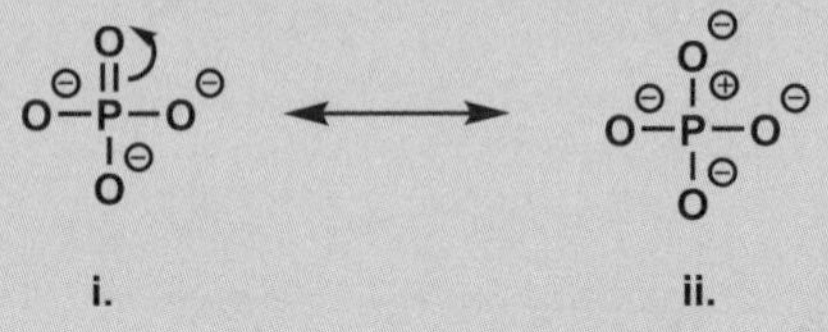

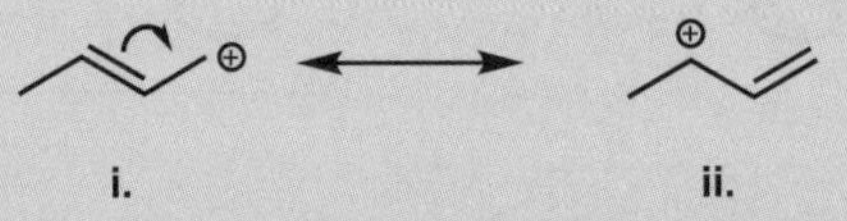

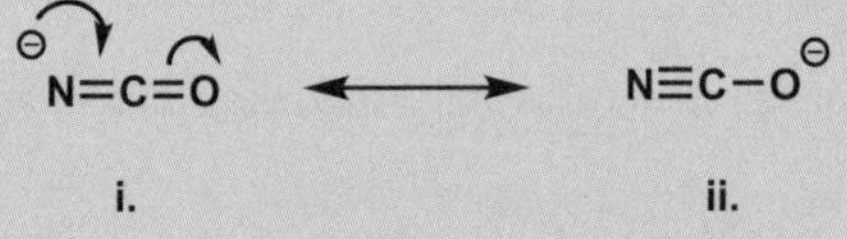

1.9 Tautomerism

Tautomerism describes the interconversion of two compounds by a shift of a double bond and hydrogen atom (or proton). This may look at first like resonance is occurring, but tautomerism actually describes an equilibrium between two distinct species that are constitutional isomers. As a result, we are able to shift the equilibrium using factors such as pH. The most commonly encountered form of tautomerism is keto-enol tautomerism, which describes the interconversion of an aldehyde or ketone possessing α-hydrogen atoms to an enol. In order to speed up this tautomerism, an acid or base catalyst can be used.

Worked example 1.9A

Draw the tautomeric forms of butanone.

Solution

First, we must draw on our knowledge of chemical nomenclature to draw out butanone. The prefix butan- means that the parent hydrocarbon chain is four carbon atoms long, while the suffix -one means that this is a ketone. Drawing out butanone based on this knowledge will give us the following structure:

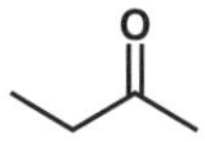

butanone

Now we need to think about what tautomers could be formed. As butanone contains a ketone with α-hydrogens, it is able to convert into an enol. However, there are α-hydrogens at both the 1 and 3 position, so there will be two possible enols formed by tautomerism. We can now draw them out by using the electrons in each C—H bond to form a π-bond with the carbon of the ketone, forming an -ene group, and moving the hydrogen proton onto the oxygen, to form an alcohol.

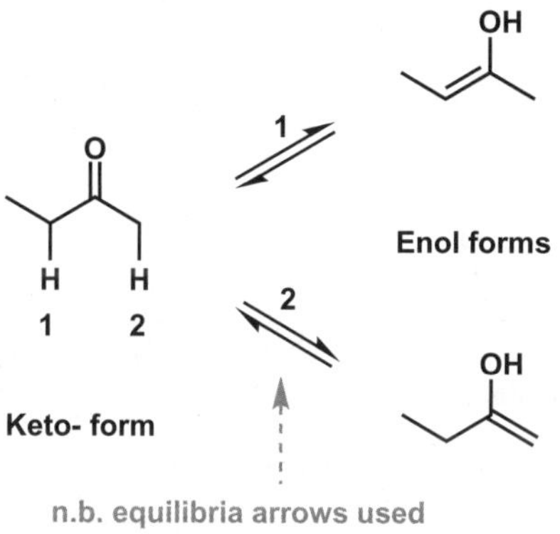

Keto- form

Enol forms

n.b. equilibria arrows used

Worked example 1.9B

The use of a base can catalyse the tautomerization of acetone (propanone) to its enol form. Provide a mechanism for this transformation.

acetone

Solution

This is one of the first mechanisms we've had to draw in this workbook, so we shall try to take care. First, the base will extract an α-hydrogen by donating a pair of electrons into the C—H σ* anti-bonding orbital, which breaks the C—H σ-bond, and forms a bond between the hydrogen and base (1). The pair of electrons in the C—H σ-bond are now able to form a π-bond with the carbon of the ketone by donating a pair of electrons into the π* orbital of the C=O bond (2). This results in the formation of an -ene group. The π-bond of the carbonyl has been broken, and the pair of electrons must reside on the oxygen atom, so that the carbon does not exceed the octet rule (3). We can now draw this using curly arrows:

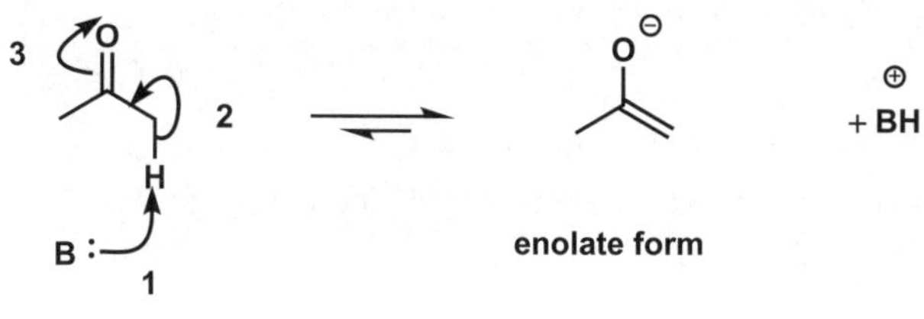

enolate form   + BH

B = Base

This has formed the enolate, which is electronically identical to the enol, but is missing a hydrogen nucleus (proton) to convert it into the enol. Some tautomerizations only proceed to this enolate, but we have been asked to draw the enol, so we shall assume that the conditions are such that a proton could be extracted from either the protonated base, or solvent.

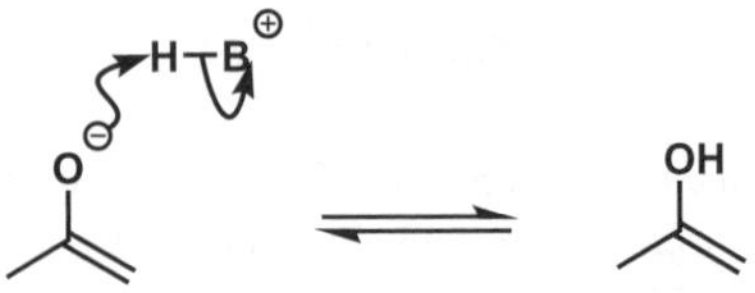

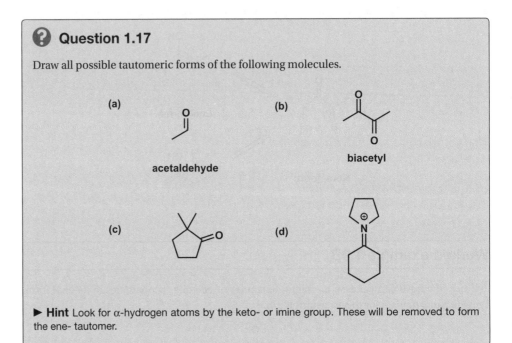

Question 1.17

Draw all possible tautomeric forms of the following molecules.

(a) acetaldehyde

(b) biacetyl

(c) 2,2-dimethylcyclopentanone

(d) cyclohexylidene pyrrolidinium

▶ **Hint** Look for α-hydrogen atoms by the keto- or imine group. These will be removed to form the ene- tautomer.

Question 1.18

Provide a mechanism for the acid catalysed tautomerization of acetone.

1.10 Synoptic questions

Question 1.19

3,4,4,5-Tetramethylcyclohexa-2,5-dienone has the trivial name 'penguinone', due to its resemblance to the flightless seabird which shares its name.

(a) Draw penguinone.
(b) Identify whether penguinone is aromatic or not.
(c) Draw all possible resonance structures of penguinone.

Question 1.20

'Michael addition' is a useful way of forming C—C bonds between a nucleophile and an α,β-unsaturated carbonyl compound. An example of a Michael addition is shown below.

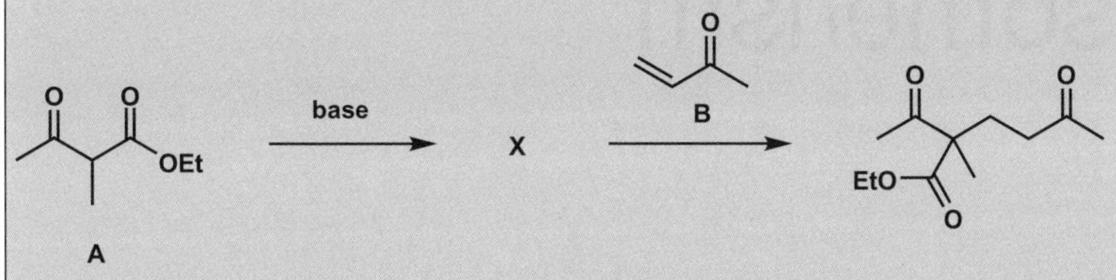

Compound 'A' can undergo Michael addition with the α,β-unsaturated ketone 'B' if it is first treated with a base to generate the reactive species X. Draw the structure of X and show any resonance forms.

References

Burrows, A., Holman, J., Parsons, A., Pilling, G., and Price, G. (2013) *Chemistry³*, 2nd edn (Oxford University Press, Oxford).

Clayden, J., Greeves, N., and Warren, S. (2012) *Organic Chemistry*, 2nd edn (Oxford University Press, Oxford).

2 Isomerism

2.1 What is isomerism?

Isomerism describes the different ways that you can arrange the atoms in a molecule with a fixed formula. This arrangement of atoms can take the form of constitutional (structural) isomerism, where the atoms are joined in a different order, or stereoisomerism, where the atoms are arranged differently in space. Stereoisomerism is subdivided into conformational isomerism and configurational isomerism. Two conformational isomers can be converted into each other (interconverted) by rotation around single bonds, whereas configurational isomers cannot be interconverted without breaking a bond.

→ Conformational isomerism is not included in this workbook, but extra information can be found in Chapter 18 of Burrows et al. (2013).

2.2 Constitutional isomerism

Two molecules that share the same molecular formula but have atoms bonded in a different order are called constitutional isomers. For instance, if we take butan-1-ol, 2-methylpropan-1-ol, 2-methylpropan-2-ol, and 1-methoxypropane, whose structures are shown in Figure 2.1, and count the number of carbon, hydrogen, and oxygen atoms in these molecules, we obtain the same molecular formula, $C_4H_{10}O$. However, it's clear that these molecules are not identical, and even possess different functional groups. Constitutional isomers can have different connections in their carbon chains (chain isomerism), different placement of identical functional groups on that carbon skeleton (positional isomerism), or different functional groups entirely (functional group isomerism). The relationship between the aforementioned butan-1-ol isomers is shown in Figure 2.1.

→ Note that two constitutional isomers may be a combination of chain, positional, and functional group isomers.

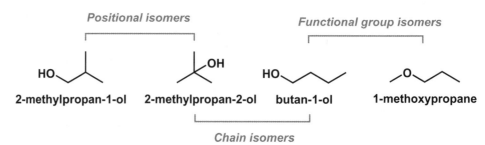

Figure 2.1 Relationships between positional isomers, chain isomers, and functional group isomers.

2.2 CONSTITUTIONAL ISOMERISM

Worked example 2.1A

3-methylbutanal, also known as isovaleraldehyde, is commonly used in the production of pesticides. Which of the compounds below are constitutional isomers of 3-methylbutanal?

3-methylbutanal

pent-1-en-3-ol

3-methylbutan-1-ol

3-methylbut-2-enal

2-ethoxypropane

butanal

2-methyltetrahydrofuran

Solution

First of all, it is necessary to count the number of carbon, oxygen, and hydrogen atoms present in 3-methylbutanal. It may help to redraw the molecule with all atoms included. This allows identification of the molecular formula of 3-methylbutanal as $C_5H_{10}O$.

C: 5
H: 10
O: 1

Molecular formula: $C_5H_{10}O$

This question can be answered by simply counting the numbers of each atom present in isovaleraldehyde, and comparing the molecular formula to each other molecule. However, by first identifying the number of double bond equivalents in 3-methylbutanal, we can identify constitutional isomers quickly: 3-methylbutanal contains one double bond, on the aldehyde functional group, so the constitutional isomers of 3-methylbutanal will contain one double bond equivalent. Only three molecules of the six contain one double bond equivalent, namely pent-1-en-3-ol, butanal, and 2-methyltetrahydrofuran.

The number of carbon and heteroatoms may then be counted to identify the correct constitutional isomers. Butanal can then be dismissed as it contains only four carbon atoms, leaving pent-1-en-3-ol and 2-methyltetrahydrofuran as the constitutional isomers of 3-methylbutanal.

→ Double bond equivalents = $C - (H/2) + (N/2) + 1$, where C = number of carbon atoms, H = number of hydrogen atoms and halogen atoms, and N = number of nitrogen atoms. See Chapter 1, section 1.5 for further help.

Worked example 2.1B

For compounds A–F, identify which pairs of compounds are isomers, then assign whether they are chain, positional, or functional group isomers of each other.

A

B

C

D

E

F

Solution

First, the molecular formula of compounds A–F must be identified by counting the number of carbon, hydrogen, and nitrogen atoms in each molecule. Those compounds with identical molecular formulae must be isomers of each other. This allows the identification of the pairs of constitutional isomers: **A** and **E** ($C_7H_{15}N$), **B** and **F** ($C_6H_{13}N$), and **C** and **D** ($C_8H_{17}N$).

Now, the pairs of isomers can be compared, and the nature of their isomerism identified. **A** contains a branched carbon-chain, whilst **E** has a straight carbon side-chain, therefore they are **chain isomers**. **B** has the methyl group at the 3-position on the cyclic piperidine ring, whilst in **F** it is bonded at the 2-position. The difference in the position of these groups makes the two molecules **positional isomers**. Finally, **C** contains a primary amine functional group, whereas in molecule **D** the nitrogen atom is within the secondary amine group. These two molecules are therefore **functional group isomers**.

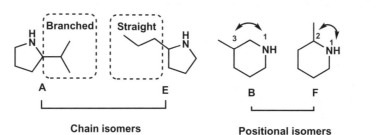
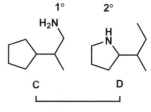

> **Question 2.1**
>
> Draw and name all six possible constitutional isomers of 1-chloropentane ($C_5H_{11}Cl$).
>
>
>
> 1-chloropentane

> **Question 2.2**
>
> Identify whether the following pairs of molecules are chain, positional, functional group isomers, or a combination of the three. Think carefully about the definitions of each!
>
> (a) and
>
> (b) and
>
> (c) and
>
> (d) and

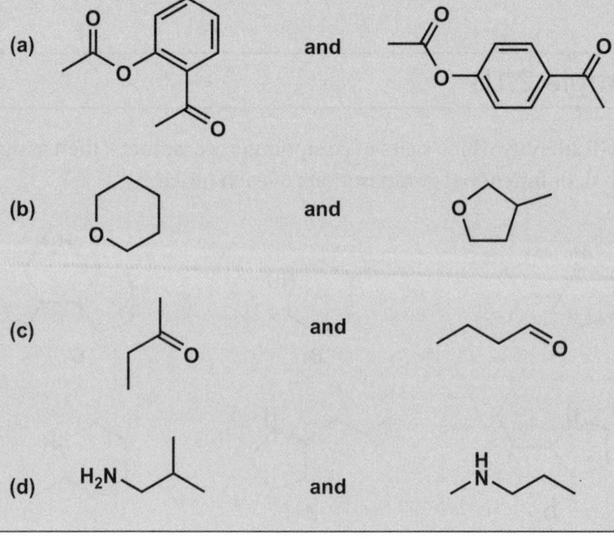

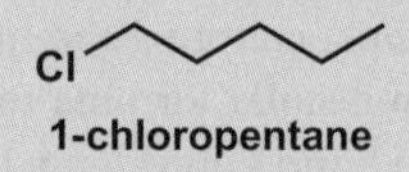

2.3 Configurational isomerism

Configurational isomers are a type of stereoisomers, so they have the same molecular formula **and order of atoms**, but have those atoms arranged differently in space. Configurational isomers cannot be interconverted without having to break at least one chemical bond. Configurational isomerism is further subdivided into *cis/trans* isomerism (*E/Z* isomerism) and optical isomerism (chirality), shown in Figure 2.2.

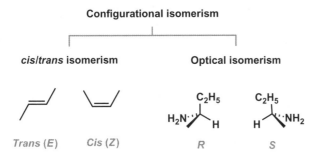

Figure 2.2 The term 'configurational isomerism' covers both *cis/trans* isomerism and optical isomerism, examples of which are shown.

Assigning the stereochemical configuration of isomers: the Cahn–Ingold–Prelog system

In order to assign the stereochemical configuration of isomers, you will first need to be able to prioritize groups according to the Cahn-Ingold-Prelog (CIP) system. Making sure that you are familiar with this system will greatly speed-up your problem solving. The CIP system allocates each group around a stereocentre a number (priority), with the relative positions of these numbers allowing the assignment of stereochemical configuration. Briefly, the CIP system is applied as follows:

1. The atomic number of atoms closest to the stereocentre determines priority, with the largest given highest priority, and so on.

e.g.

Order of priority by the CIP system:	R—SH	>	R—OH	>	R—NH$_2$	>	R—CH$_3$	>	R—H
Atomic number of first bonded atom:	16 (S)		8 (O)		7 (N)		6 (C)		1 (H)

2. If the atoms have the same atomic number, then the atomic number of the next atom away from the stereocentre is examined. If no difference is seen then continue to move away from the stereocentre until a difference is found.

e.g.

Order of priority by the CIP system: R—CH$_2$OH > R—CH$_2$CH$_2$— > R—CH$_2$— > R—CH$_3$

3. If the substituent contains a double or triple bond, then it is treated as being bonded to that atom twice, or three times, respectively, e.g. C(H)=CH$_2$ is prioritized over CH$_2$CH$_3$.

e.g.

Order of priority by the CIP system: R—C≡ > R—CH=CH— > R—CH$_2$—

Furthermore: R—C(=O)OH > R—C(=O)H > R—CH$_2$OH

→ In the rare case that you might be assigning priority to two different isotopes of the same element (which would therefore have the same atomic number) you give priority to the atom with the highest atomic weight. For example, H and D have the same atomic number, but D has a larger atomic weight. So in this case D would be prioritized over H.

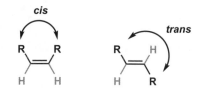

Figure 2.3 Example *cis* and *trans* isomers of a general alkene.

➡ The (Z) and (E) notations are taken from German—Zusammen (Z) means 'together', and Entgegen (E) means 'opposite'.

2.4 *Cis/trans* isomerism

Cis/trans isomers arise when rotation is restricted at a particular point in a molecule, most commonly due to a carbon–carbon double bond. This restricted rotation allows groups positioned either side of the double bond to either sit adjacent to each other (*cis*), or opposite each other (*trans*), as shown in Figure 2.3. For disubstituted alkenes, the *cis/trans* system can easily be applied by eye. However, for tri- and tetra-substituted alkenes, the relative priority of the two substituents on each side of the stereocentre must be assigned using CIP rules. When this system is used, the isomers are assigned the letters (Z) or (E), instead of *cis* or *trans*, respectively.

To assign *E/Z* isomers you need to look for the relative positions of the highest priority groups on either side of the double bond. If the highest priority groups either side of the double bond are adjacent to each other, the isomer is (Z) and vice-versa.

Worked example 2.2A

Identify whether the following three molecules are *cis* or *trans* isomers.

Solution

(a) We need to look carefully at the relative position of the ethyl and methyl groups either side of the double bond. Rotation cannot occur around this double bond, and the structure provided shows us that these groups sit on opposite sides of the double bond. This molecule is therefore the *trans* isomer of pent-3-ene.

(b) It is important with any kind of stereoisomer firstly to identify the point(s) at which the stereoisomerism arises (the stereocentre). In this example, there are two double bonds, and therefore two points in the molecule at which no rotation is possible. The carbonyl group (C=O) does not have any substituents attached to the oxygen atom, so cannot give rise to *cis/trans* isomers. However, the carbon–carbon double bond has two different substituents attached to either carbon atom, and can give rise to *cis/trans* isomers. This is the stereocentre we need to assign. Looking at the double bond, we can see that the methyl group on one side of the double bond and the carbonyl group on the other are adjacent to each other. This is, therefore, the *cis* isomer.

(c) This question requires us to look at the *cis/trans* isomerism arising from restricted rotation around a ring, but we follow the same principles as for double bonds to assign which isomer is present. The two amine groups attached to the cyclohexane ring are on opposite

2.4 CIS/TRANS ISOMERISM

sides of the ring to each other—one sits 'up' from the plane (the wedge), the other 'down' (the dashes). This means that this is the *trans* isomer.

Worked example 2.2B

For the following molecules, identify whether they are *E* or *Z* isomers.

(a) (b) (c)

Solution

(a) For the left-hand side of this molecule, assigning the priority of the substituents is easy—hydrogen has the lowest atomic number of any element, so will be lower priority than any other element, according to the CIP rules. The right-hand side of the molecule looks slightly more difficult. But as with all questions regarding the assignment of priority, assignment is simple if the CIP rules are systematically applied. Here, we have to assign priority to —OH or —CH$_3$. Oxygen has an atomic number of eight compared to carbon's six, and therefore takes priority. This puts the two groups with the highest priority adjacent to each other—this is, therefore, a (*Z*) isomer.

➔ This molecule may have looked as if it were *trans* (*E*), because the methyl groups were on opposite sides, but remember that the stereochemistry is assigned based on the priority of the groups using CIP, rather than their similarity.

(b) Assignment of the left-hand side of this molecule requires the use of point 2 of the CIP rules outlined previously. As we travel away from the stereocentre, the first atom met is carbon in both cases. We need to move further along bonds in order to assign priority. One of the substituents is a methyl group, with a carbon atom bonded to three hydrogen atoms. The other is an ethyl group, and bears a carbon atom bonded to two hydrogen atoms and one carbon atom. Carbon has a higher atomic number than hydrogen, so the ethyl group takes priority. The right-hand side of this molecule is simply assigned based on atomic number, with the carbon atom of the ethyl group taking priority over the hydrogen atom. The highest priority groups are adjacent to each other, so this is the (*Z*) isomer.

(c) The left-hand side of this molecule requires us to assign priority to either an ethyl group or an ethynyl substituent. Although both of the closest carbon atoms are bonded to an additional carbon atom, the ethynyl group has a carbon–carbon triple bond, so is treated as if it were bonded to three carbon atoms, rather than the single carbon in the ethyl group. The ethynyl group is therefore higher priority, according to the CIP rules. The amine group on the right-hand side has a higher priority as nitrogen has a higher atomic number than carbon. The two highest priority groups are opposite each other, so this is the (*E*) isomer.

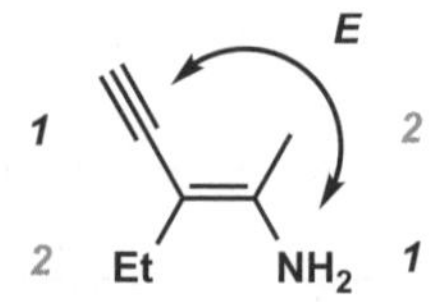

> ### Question 2.3
>
> Rank the following groups in order of their priority according to the CIP rules.
>
> (a) —H   —NH$_2$ —CH$_3$ —OH
>
> (b) (propenyl)   (ethynyl)   —CH$_3$ —H
>
> (c) —CH$_3$ —Et   —H   —NH$_2$
>
> (d) —NH$_2$ —CN   —NMe$_2$ —NO$_2$
>
> ▶ **Hint** Ensure you are familiar with the CIP system before you begin, and systematically apply them in the order shown in section 2.3. Extra information is available in Burrows et al. (2013) and Clayden et al. (2012) if you need help.

> ### Question 2.4
>
> Identify if the following molecules are *cis* or *trans* isomers.
>
> (a) 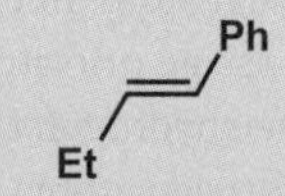 Et/Ph alkene   (b) 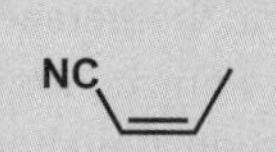 NC alkene   (c) 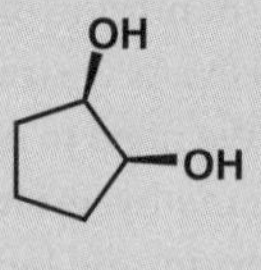 cyclopentane-1,2-diol   (d) 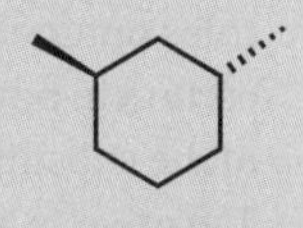 disubstituted cyclohexane
>
> ▶ **Hint** In disubstituted rings, the relative positions of the two substituents does not affect whether an isomer is *cis* or *trans*. Just treat them as you would a 1,2-disubstituted system.

Question 2.5

Use the CIP system to assign the following molecules as either (*E*) or (*Z*) isomers.

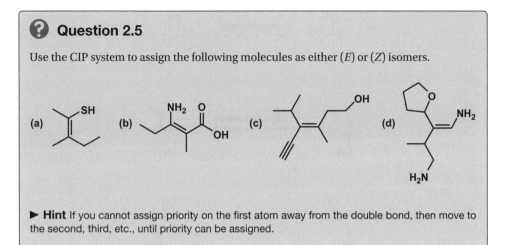

▶ **Hint** If you cannot assign priority on the first atom away from the double bond, then move to the second, third, etc., until priority can be assigned.

2.5 Optical isomerism (chirality)

A molecule with a non-superimposable mirror image is said to be chiral. In the majority of cases that you will encounter in organic chemistry, this chirality arises at a carbon atom bonded to four **different** chemical groups. This type of isomerism is often termed optical isomerism as these molecules have the ability to rotate plane-polarized light. Figure 2.4 shows a chiral compound, X, and its mirror image, Y. If we imagine turning Y 180° horizontally, we get out a structure which is not superimposable on the original molecule, X. X is, therefore, chiral.

Figure 2.4 X is a chiral molecule, as its mirror image is not superimposable on itself.

The two chiral stereoisomers arising from a single stereocentre are designated as either (*R*) or (*S*). In order to assign chiral molecules as either (*R*) or (*S*), you must firstly prioritize the substituents around the stereocentre according to the CIP system. Then, the molecule must be rotated so that the lowest priority substituent (most commonly hydrogen) is facing away from the viewer. Now, if the priority of the substituents decreases clockwise around the molecule, the isomer is (*R*), if the priority decreases anticlockwise, then it is (*S*).

→ You may see optical isomers designated as (+)- and (−)- or *d*- and *l*-. These prefixes pertain to the direction that plane polarized light is rotated, and are not covered in this workbook. You may also see molecules assigned as D- or L-, which is a slightly more complex method of assigning chirality (usually to amino acids) based on a separate set of rules called the CORN rules.

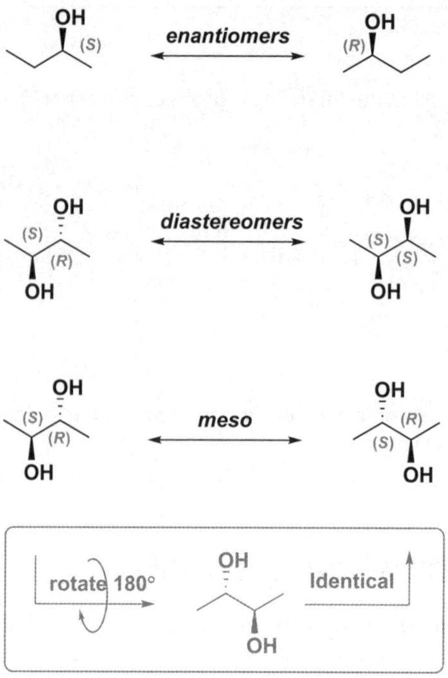

Figure 2.5 The relationship between enantiomers, diastereoisomers, and meso compounds.

→ Note that inverting all the stereocenters in a chiral molecule gives the enantiomer, unless the compound is meso. By contrast, inverting some, but not all, of the stereocenters gives a diastereomer.

A chiral molecule can also exist as a non-superimposable mirror image of itself: an enantiomer. However, if a chiral molecule contains more than one stereocentre giving rise to chirality, it can be related to stereoisomers with different (R)/(S) configurations that are not mirror images. These compounds are said to be diastereomers (or diastereoisomers). In specific cases, a diastereomer can have a configuration such that it **is** imposable on its own mirror image, and is therefore not chiral! In this case it is said to be a meso compound. The relationship between chiral isomers is shown in Figure 2.5.

Worked example 2.3A

Assign the following chiral molecules as either the (R) or (S) enantiomers.

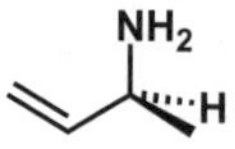

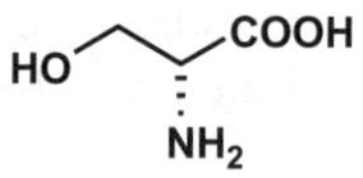

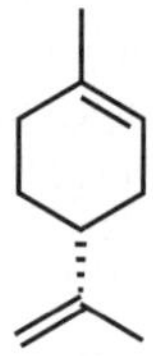

Solution

(a) If we follow the same logic used in the explanatory paragraph at the start of this section, then we will always arrive at the correct answer. First, we must assign priority to the substituents using the CIP system. This puts the amine group as the highest priority, the ethenyl group second, the methyl group third, and the hydrogen atom fourth. If you require more practice assigning priority using the CIP system, then revisit section 2.4. The lowest priority substituent, i.e. the hydrogen atom, is already facing away from the viewer, so no rotation of the molecule is necessary. The substituents are now ordered in an anticlockwise fashion, so the isomer is (S).

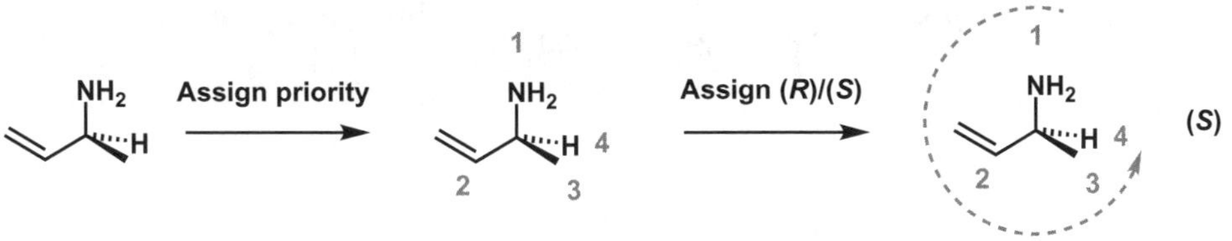

(b) In this molecule, the hydrogen atom has not been drawn out for you, but it is still there. However, you may find it helpful to redraw the structure with the atom included. Assigning priority using the CIP system puts the groups in the following order:

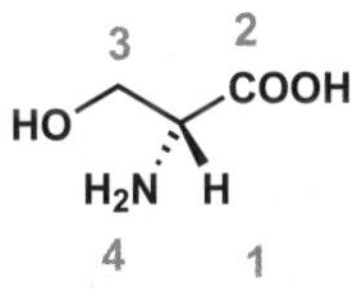

Before we are able to assign whether this stereoisomer is R or S, we need to make sure the lowest priority group is facing away from us. Rotating the molecule 180° horizontally, then making an assignment allows us to identify this as the R enantiomer.

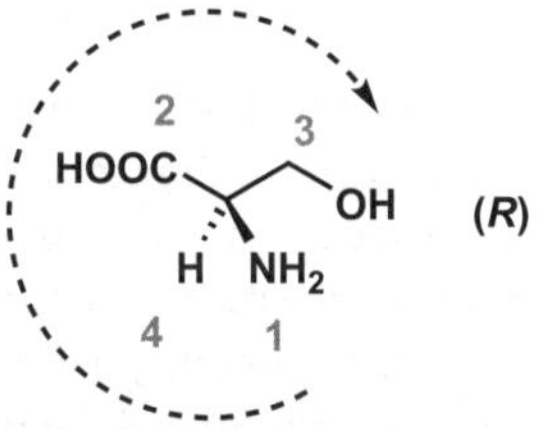

(c) This molecule may look a bit different from those we have previously assigned, but there is still a stereocentre which gives rise to chirality in the molecule. Assigning priority is a little difficult, due to the ring, but it is still possible if we follow the CIP system. The assignment of priority must be done with careful consideration of the relative positions of the C—C double bonds. In this case, hydrogen is lowest priority, followed by the left-hand side of the ring, then the right, and finally the C(=CH$_2$)CH$_3$ substituent. The lowest priority group is facing the viewer, so the molecule is rotated 180° horizontally. The order of priority now decreases in a clockwise fashion, so the molecule is the (R) enantiomer.

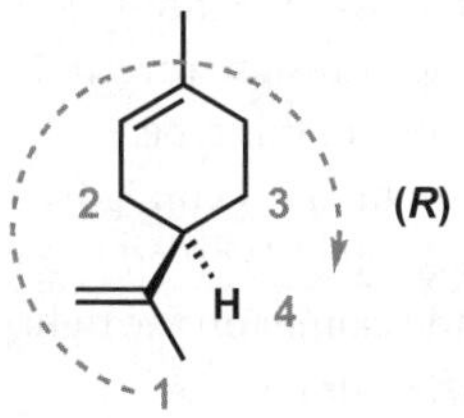

Worked example 2.3B

Are the following pairs of stereoisomers enantiomers, diastereomers or meso?

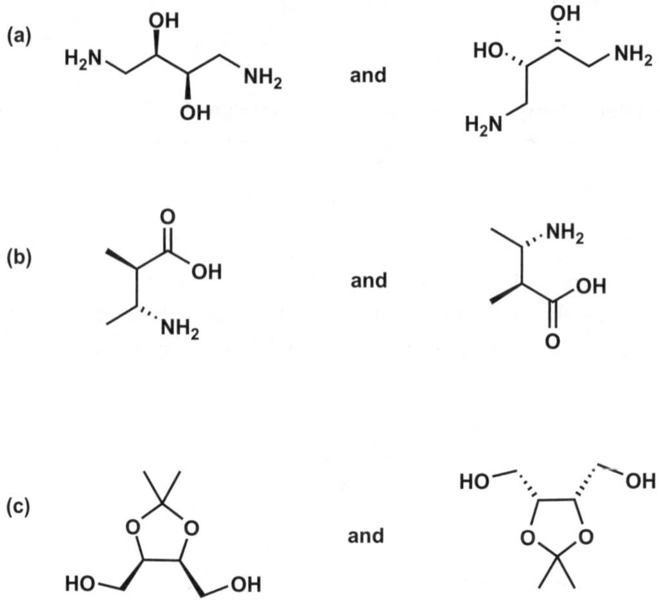

(a) [structure] and [structure]

(b) [structure] and [structure]

(c) [structure] and [structure]

Solution

(a) This set of questions can be answered using two different methods. Either you can redraw the molecules in order to see if they are mirror images of each other, or you can assign the stereochemical configuration and work out their relationship from there. If we redraw the molecule on the right-hand side so that the carbon backbone mirrors the molecule on the left, we find that these isomers are not mirror images of each other due to the chiral centres, but are still non-superimposable. This means that the molecules are diastereomers.

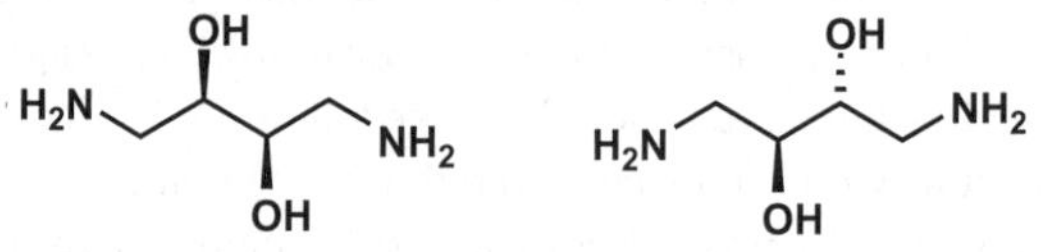

[structures]

You may have difficulty rearranging molecules in this fashion, particularly as the complexity of the molecules you are trying to assign increases. Another method of working out relationships between these stereoisomers is to assign the stereochemical configuration to each chiral centre first. If all stereocentres are inverted, the isomers are enantiomers; if some—but not all—of the stereocentres are inverted, they are diastereomers. This will always be a good strategy in questions of this type.

First of all, we assign the stereochemical configuration to the two chiral centres on each isomer using the CIP system covered in Worked example 2.6A. This gives us (R,R) for the isomer on the left, and (S,R) for the isomer on the right. As only one of the two stereocentres has been inverted, these are diastereomers.

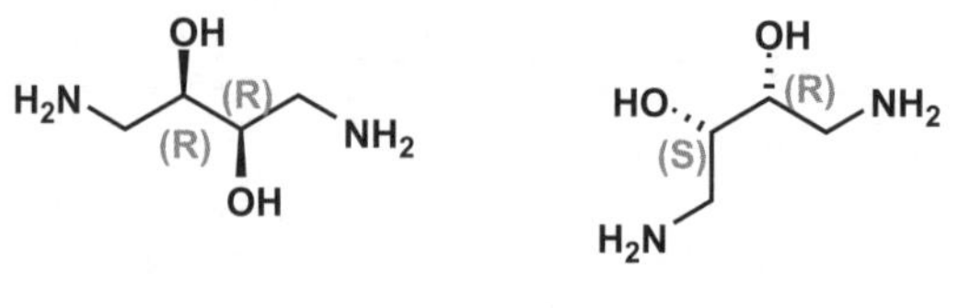

[structures]

2.5 OPTICAL ISOMERISM (CHIRALITY)

(b) Firstly, assign the stereochemical configuration of the two isomers shown in this question. This gives us (R,R) for the left-hand molecule, and (S,S) for the right. As both stereocentres have been inverted, these are enantiomers. Furthermore, simply rotating the molecule on the right hand side 180° anticlockwise gives us a non-superimposable mirror image of the compound on the left.

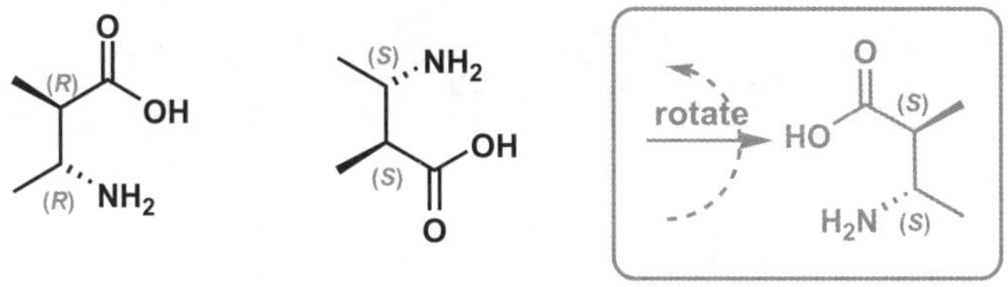

(c) Again, assign the stereochemical configuration of each stereocentre first. Notice how the configuration of each stereocentre has not changed, giving (R,S) in each instance. The molecule is, therefore, meso. If we rotate the molecules 180° vertically, they are superimposable.

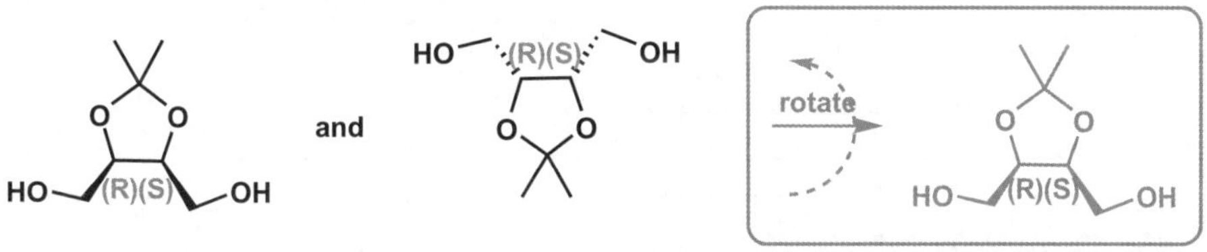

For future questions, you can use a quick trick to help you identify meso compounds: they all contain at least one plane of symmetry (Figure 2.6). This plane of symmetry is what allows them to be superimposable on their mirror image. Remember that C—C bonds can rotate, barring steric effects, so sometimes you may have to change the orientation of a molecule to notice this symmetry.

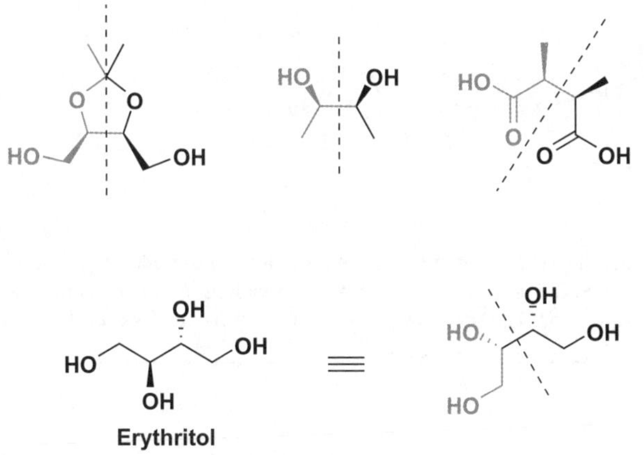

Figure 2.6 Symmetry in meso compounds.

Question 2.6

Find and mark all chiral centres in the following molecules.

(a) 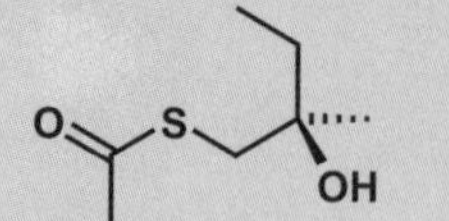 [structure showing acetyl-S-CH2-C*(OH)(Et) with wedge bonds]

(b) [cyclopentene with isopropyl and OH substituents]

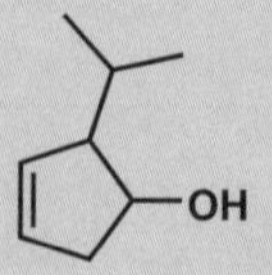

(c) [Ph-CH2-CH(Br)-CH(CH3)2 type structure with Br]

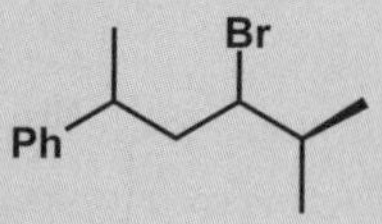

(d) 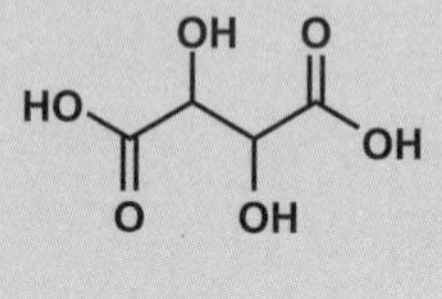 HO-C(=O)-CH(OH)-CH(OH)-C(=O)-OH (tartaric acid)

(e) 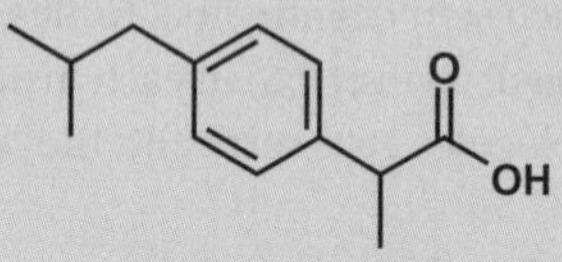 [ibuprofen structure: isobutyl-phenyl-CH(CH3)-COOH]

(f) 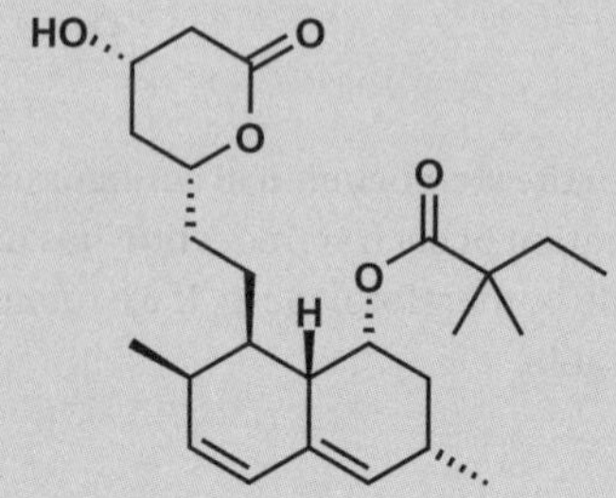 [simvastatin-like structure with lactone, decalin, and ester groups]

▶ **Hint** Remember that for a carbon atom to be chiral it must be bonded to four **different** groups.

Question 2.7

Assign the following molecules as either the (R) or (S) isomers.

(a) 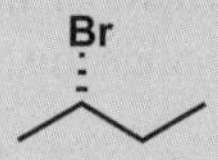 [CH3-CH(Br)-CH2-CH3 with wedge Br]

(b) 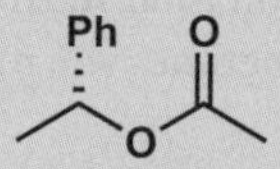 [Ph-CH(CH3)-O-C(=O)-CH3]

(c) [HC≡C-C*(SiMe3)-CH2-NH2]

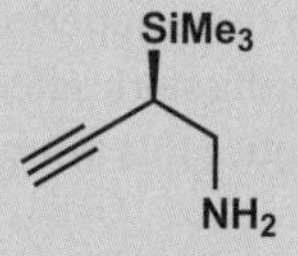

(d) 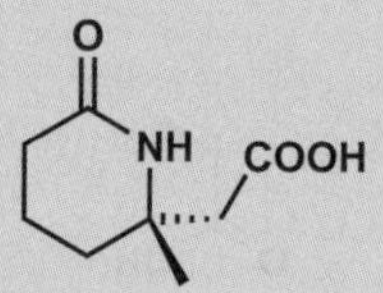 [δ-lactam with methyl and CH2COOH substituents]

(e)  [epoxide with tBu and Et substituents]

(f) 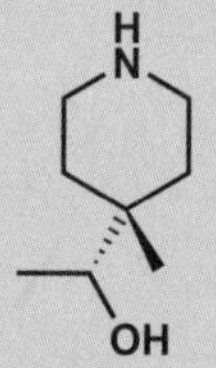 [4-methylpiperidine with CH2OH substituent]

▶ **Hint** If you find it difficult to rotate the molecule so that the lowest priority group is facing away from you, then leave is where it is, assign the stereochemical configuration, and reverse the answer at the end (i.e. (R) goes to (S), (S) goes to (R))—you'll end up with the correct isomer that way!

Question 2.8

Are the following pairs of compounds enantiomers, diastereomers, meso, or not stereoisomers at all?

(a) [structure: Ph-C(OH)(CH₃)-CH₂CH₃ with wedge/dash] and [structure: Ph-C(OH)(CH₃)-CH₂CH₃ with opposite wedge/dash]

(b) [structure: 2-methyl-4-hydroxypyrrolidine] and [structure: 2-hydroxy-4-methylpyrrolidine]

(c) [structure: HS-CH₂-CH(CH₃)-CH(CH₃)-CH₂-SH] and [structure: HS-CH₂-CH(CH₃)-CH(CH₃)-CH₂-SH]

(d) [structure: 4-methylcyclohex-2-enol] and [structure: 4-methylcyclohex-2-enol]

(e) [structure: 1-phenyl-2-(methylamino)propan-1-ol] and [structure: 1-phenyl-2-(methylamino)propan-1-ol]

(f) [structure: methyl 2-phenyl-2-(piperidin-2-yl)acetate] and [structure: methyl 2-phenyl-2-(piperidin-2-yl)acetate]

▶ **Hint** If you are confident that you can rotate these compounds to try and mirror each other, you should reach an answer. If you are unsure, however, assigning the stereochemical configuration of the chiral centres will lead you to a correct answer.

2.6 Synoptic questions

Question 2.9

(a) You are provided with an unknown colourless liquid. Elemental analysis shows that this sample has the molecular formula C_4H_9Br. It is your task to identify this compound. Firstly, draw all possible constitutional isomers of this sample.

(b) You now place this sample into a polarimeter. You find that the optical rotation equals +4.5°. Draw and name the two compounds that this sample could be.

(c) Treating these compounds with a strong base can lead to an E2 elimination of the bromine atom. This reaction could also potentially eliminate three different hydrogen atoms, leading to three different products. Draw and name these products.

▶ **Hint** In (a) calculating the number of double bond equivalents will help! In (c) you may struggle with this question if you have not covered elimination reactions before. Don't worry if you haven't yet!

Question 2.10

Pseudoephedrine is a commercially available decongestant. You are provided with a sample of the dextrorotatory (+) enantiomer of pseudoephedrine, shown below. You measure the optical activity of this sample and find that (+)-pseudoephedrine has a specific rotation of +52°.

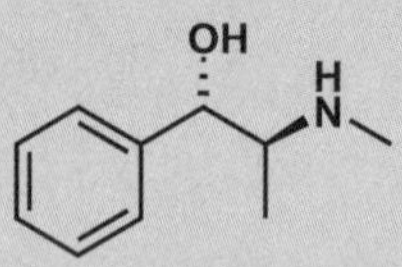

(a) Assign the stereochemical configuration of the two stereogenic centres of (+)-pseudoephedrine.

(b) You obtain a different, unknown, stereoisomer of pseudoephedrine, 'X', and measure its optical activity. You find that this stereoisomer has a specific rotation of −52°. What is the relationship of X to (+)-pseudoephedrine?

(c) Draw the structure of X.

References

Burrows, A., Holman, J., Parsons, A., Pilling, G., and Price, G. (2013) *Chemistry*[3], 2nd edn (Oxford University Press, Oxford).

Clayden, J., Greeves, N., and Warren, S. (2012) *Organic Chemistry*, 2nd edn (Oxford University Press, Oxford).

3
Nucleophilic substitution

3.1 Electrophiles and nucleophiles

What is an electrophile?

An electrophile is a neutral or positively charged species with an empty orbital (or an energetically accessible anti-bonding orbital) which can accept electrons. Lewis acids, e.g. $AlCl_3$ and BCl_3, can also be considered electrophiles as they have an empty orbital that can accept an electron pair.

→ For more information on Lewis acids and bases, see section 3.2.

→ A curly arrow denotes the movement of electrons.

What is a nucleophile?

A nucleophile contains a pair of electrons that can be used to form a new chemical bond. Nucleophiles act as electron donors. When we draw a reaction mechanism we always draw the electrons flowing from the nucleophile to the electrophile.

→ The electrons that are donated can either be in the form of a lone pair or a formal negative charge.

Worked example 3.1A

In the reaction below, which molecule is the electrophile and which molecule is the nucleophile?

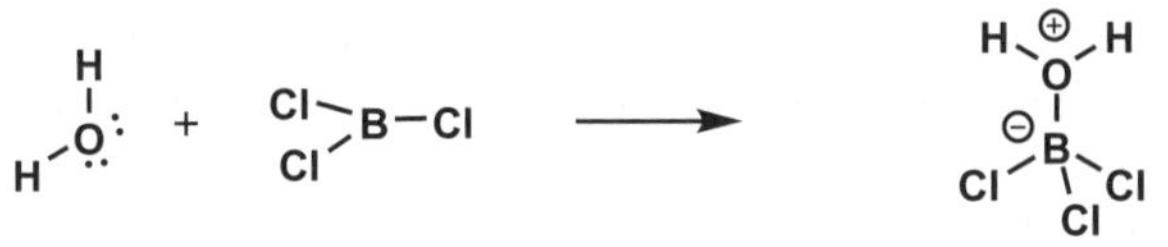

Solution

The water molecule contains two lone pairs of electrons which can be donated. Boron trichloride contains an empty p orbital perpendicular to the plane of the molecule, which can accept electrons. If we look at the product, the oxygen has become positively charged and therefore must have donated a pair of electrons. By contrast, the boron has become negatively charged and therefore must have accepted a pair of electrons. This means that in this case water is a nucleophile, as it has donated a pair of electrons, and the boron species is an electrophile, as it has accepted a pair of electrons.

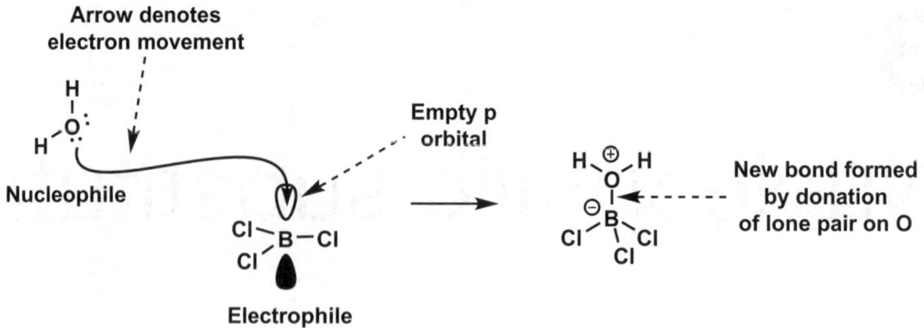

3 NUCLEOPHILIC SUBSTITUTION

Worked example 3.1B

In the reaction below, which molecule is the electrophile and which molecule is the nucleophile?

Solution

If we look at the reaction carefully we can see that the tosylate and bromide have swapped positions in the products when compared to the reactants. This therefore implies that the bromine must have used its electrons to displace the tosyl group, forming a new bond between the carbon and bromine.

We also know that sodium bromide, NaBr, exists as an ionic solid and can therefore be present as Na^+ and Br^- ions. Bromide (Br^-) has a negative charge which can be donated; it is therefore likely to act as a nucleophile. Consequently, *iso*-butyl tosylate will be the electrophile.

More specifically, the C–O σ* orbital is the electrophile, as this is the lowest energy unoccupied molecular orbital (LUMO). The tosylate leaves as a negatively charged salt; formally the positively-charged sodium will be associated with it, generating the products shown.

→ A tosylate/tosyl group is a *para*-toluenesulfonyl group. It is an extremely good leaving group so is commonly used in nucleophilic substitution reactions.

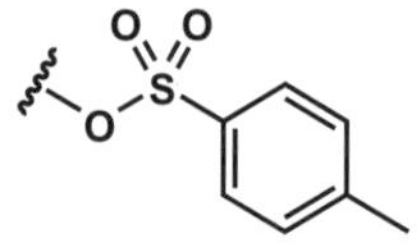

→ For more information on leaving groups, see section 3.4 and Clayden et al. (2012).

→ A nucleophile must attack an electrophile. Two electrophiles or two nucleophiles cannot react together!

→ HOMO and LUMO are discussed further in Chapter 1.

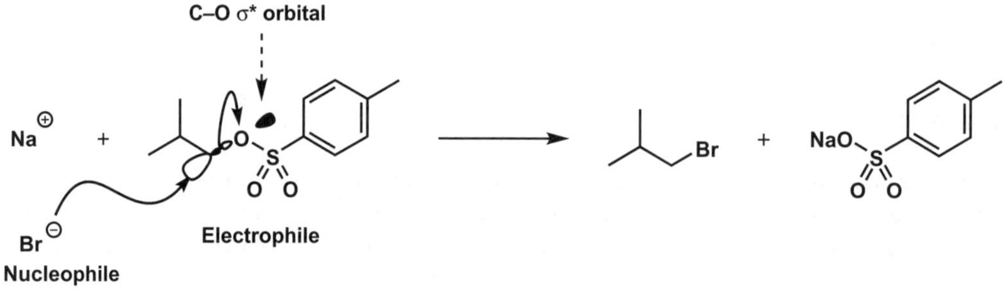

3.2 LEWIS ACIDS AND BASES

> **Question 3.1**
>
> Suggest which compound is likely to be an electrophile, and which a nucleophile, in the following series.
>
> BF$_3$ AlCl$_3$ H$_2$S   MeOH   NaOH   H$_2$O
>
> ▶ **Hint** Consider the HOMO and LUMO in each case.

> **Question 3.2**
>
> Suggest which species is likely to be the electrophile, and which the nucleophile, in the reaction below. Suggest structures for the products.
>
>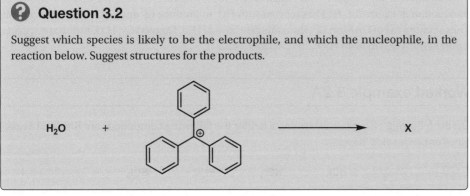

> **Question 3.3**
>
> For each step depicted, which species is the electrophile and which is the nucleophile?
>
>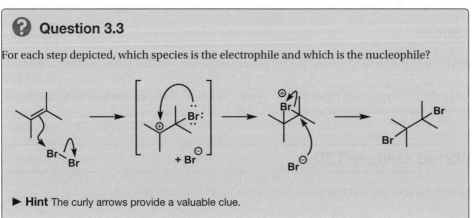
>
> ▶ **Hint** The curly arrows provide a valuable clue.

3.2 Lewis acids and bases

What is an acid and a base?

There are two different types of acids and bases: Brønsted acids/bases and Lewis acids/bases. Brønsted acids and bases are those that you have, most likely, encountered previously. A Brønsted acid is defined as a molecule that can donate a proton; a Brønsted base is a molecule that can accept a proton. Acids and bases always occur in pairs therefore if an acid is present, there will also be a conjugate base (Figure 3.1).

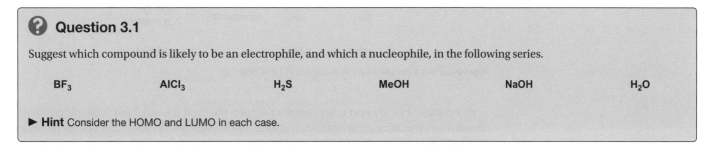

Figure 3.1 An example of a Brønsted acid and base reaction.

$$BF_3 + NH_3 \rightleftharpoons F_3B^- - N^+H_3$$

acid   base   adduct

Figure 3.2 An example of a Lewis acid and base reaction.

By contrast, a Lewis acid is an electron pair acceptor and a Lewis base is an electron pair donor. A Lewis acid can also be thought of as an electrophile and a Lewis base as a nucleophile (Figure 3.2).

As mentioned earlier, acids and bases always occur in pairs. An example of this is shown in the reaction in Figure 3.1. HCl has reacted with HO⁻ to produce Cl⁻ and H_3O^+. HCl is the acid; it loses a proton to H_2O to form Cl⁻, the conjugate base of HCl. Conversely, H_2O is the base: it gains a proton from HCl to form H_3O^+, its conjugate acid.

Worked example 3.2A

For the following examples, determine whether the following compounds are Brønsted acids/bases or Lewis acids/bases.

$$HCl \quad\quad NEt_3 \quad\quad Et_2O \quad\quad BCl_3$$

Solution

HCl is a Brønsted acid because it can act as a proton donor. NEt_3 is a Lewis base as it has a lone pair on the nitrogen which can be donated. Et_2O, diethyl ether, is a Lewis base because it has lone pairs on oxygen that it can donate. Finally, BCl_3 is a Lewis acid because it can accept an electron pair into its empty p orbital.

Worked example 3.2B

Identify the acid and conjugate base pairs for each of the following reactions.

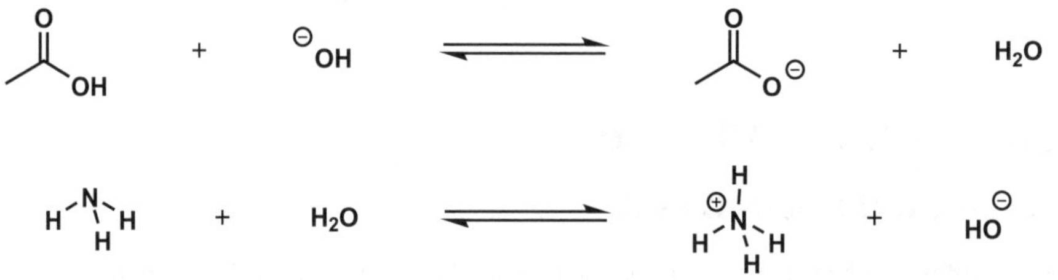

$$CH_3COOH + {}^-OH \rightleftharpoons CH_3COO^- + H_2O \quad\quad [1]$$

$$NH_3 + H_2O \rightleftharpoons NH_4^+ + {}^-OH \quad\quad [2]$$

Solution

In reaction 1, the acid is acetic acid as it loses a proton to become acetate, the conjugate base. The base is hydroxide as it gains the proton to become water, the conjugate acid.

$$\underset{\text{acid}}{CH_3COOH} + \underset{\text{base}}{{}^-OH} \rightleftharpoons \underset{\text{conjugate base}}{CH_3COO^-} + \underset{\text{conjugate acid}}{H_2O}$$

In reaction 2, ammonia is the base: it accepts a proton to form ammonium, the conjugate acid. Water is the acid: it loses a proton to form hydroxide, the conjugate base.

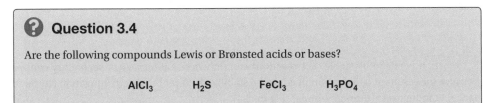

> **Question 3.4**
>
> Are the following compounds Lewis or Brønsted acids or bases?
>
> $AlCl_3$ $\quad$ H_2S $\quad$ $FeCl_3$ $\quad$ H_3PO_4

> **Question 3.5**
>
> Identify the acid/base pairs for each of the following reactions.
>
>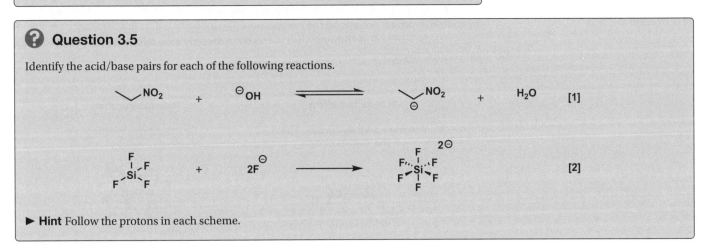
>
> ▶ **Hint** Follow the protons in each scheme.

3.3 S_N1 and S_N2

What are S_N1 and S_N2?

S_N1 and S_N2 are two classes of reaction that describe the nucleophilic substitution, or replacement, of one functional group at a **saturated carbon centre** by another. They are defined according to the molecularity (the number of species required) of the rate-determining step (RDS), or the slow step. A general scheme is shown in Figure 3.3.

➔ Nucleophilic substitution is different to aromatic substitution. Aromatic substitution is discussed in Chapter 6.

The S_N1 reaction

Rate = k_1 [substrate]

An S_N1 reaction is independent of the concentration of the incoming nucleophile and is related solely to the ability of the leaving group to leave. In an S_N1 reaction a discrete positive charge is formed on a carbon. The key point to remember is that the intermediate carbocation in an

Figure 3.3 A general substitution reaction.

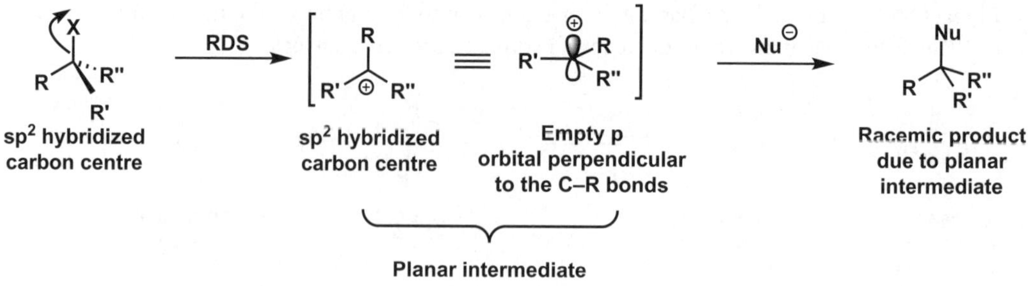

Figure 3.4 The S_N1 reaction.

S_N1 reaction needs to be stable. S_N1 reactions can only occur at tertiary or secondary centres because the resultant carbocation needs to be stabilized, either by hyperconjugation (through space) or inductively (through bonds) (Figure 3.4).

Hyperconjugation involves the overlap of σ bond (e.g. a C—H or a C—C) with the empty, hybridized p orbital present on the carbocation. The inductive effect is where electrons are attracted to the carbocation through the C—C σ bond **directly attached** to the carbocationic carbon centre, Figure 3.5.

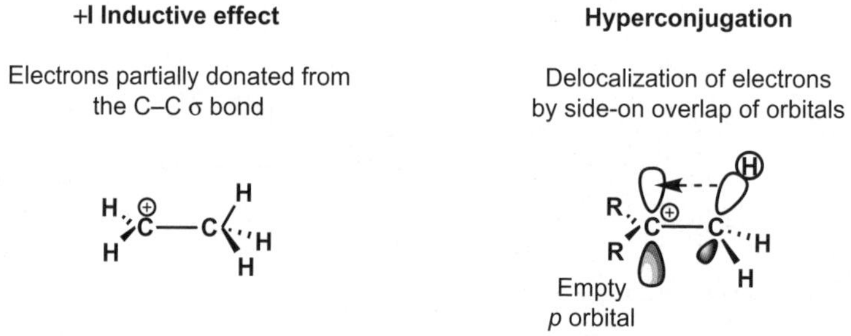

→ For a reminder about hybridization, see Chapter 1, section 1.4.

→ For more information, see Chapter 19 of Burrows et al. (2013).

Figure 3.5 The inductive effect and hyperconjugation.

Worked example 3.3A

Suggest a mechanism for the following reaction.

Solution

The first thing to notice is that the reaction is under acidic conditions so the oxygen will be protonated to give an oxonium species (a positively charged oxygen), that can then leave, generating a tertiary carbocation. This cation is attacked by the bromide anion, generating a new C—Br bond.

→ Under acidic conditions HO⁻ is never a leaving group!

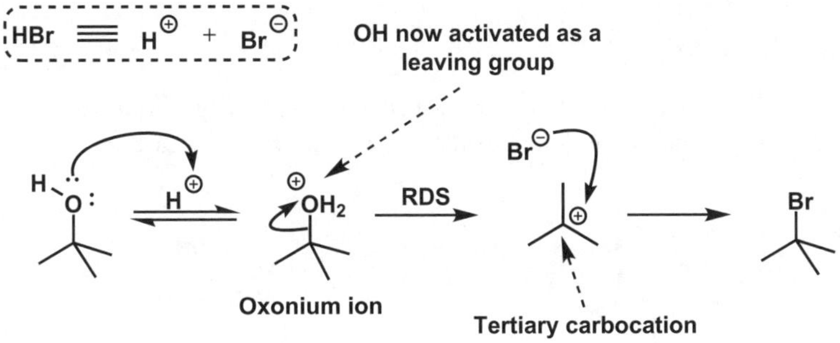

Worked example 3.3B

Draw a mechanism for the following reaction. Suggest what will happen to the stereochemical configuration of the carbon marked with an asterisk.

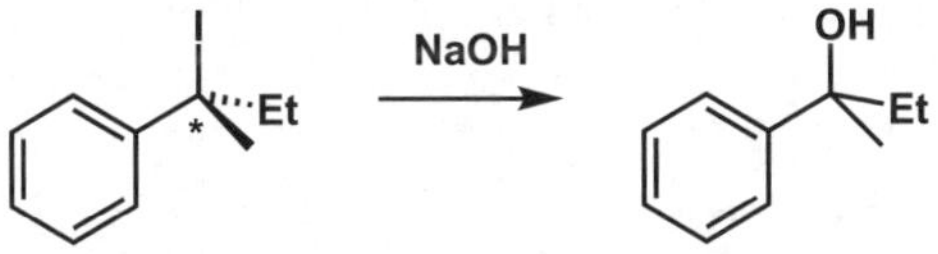

Solution

In this example, iodide is the leaving group. This iodine resides on a tertiary carbon therefore the mechanism must be S_N1, not S_N2; when the iodide leaves a stable carbocation is generated. The cation formed has sp^2 hybridization and therefore has an empty p orbital. This p orbital can be attacked by the nucleophile from either above (pathway a), or below (pathway b) the plane, leading to two different products. These products are enantiomers.

→ For more information about enantiomers and stereochemistry, see Chapter 2.

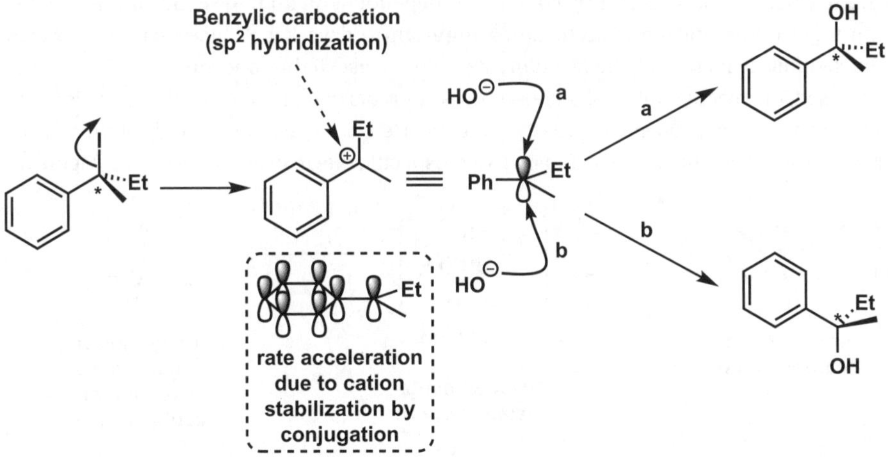

3 NUCLEOPHILIC SUBSTITUTION

> ### ❓ Question 3.6
>
> Provide a mechanism for the following transformation, assuming it undergoes an S_N1-type reaction pathway. Suggest the stereochemical configuration at the carbon labelled with an asterisk.
>
> [Ph–C*H(tBu)–Cl] → (KI, acetone, reflux) → [Ph–C*H(tBu)–I]
>
> ▶ **Hint** Draw the mechanism first and then consider the stereochemical implications.

> ### ❓ Question 3.7
>
> Provide a mechanism and product for the following transformation. Suggest the stereochemical configuration at the carbon labelled with an asterisk.
>
> [sec-butyl acetate with *iPr and Et groups, OAc wedged down] → EtSH → X
>
> ▶ **Hint** Draw the mechanism first and then consider the stereochemical implications.

The S_N2 reaction

Rate = k_2 [substrate][nucleophile]

An S_N2 reaction is a bimolecular reaction that is dependent on both the concentration of the incoming nucleophile and the ability of the leaving group to leave. If the concentration of the incoming nucleophile is increased, the reaction rate will increase; if the concentration of the substrate in increased, the reaction rate will increase. The key points to remember are that there is no intermediate in an S_N2 reaction and the reaction is concerted, S_N2 reactions can only occur at primary or secondary centres and S_N2 reactions are stereospecific due to orbital constraints (Figure 3.6).

⮕ For a full orbital explanation of this reaction, see Clayden et al. (2012).

[Nu⁻ + R'C(R)(H)X → [Nu···C···X]‡ transition state (pseudo pentavalent carbon) → Nu–C(R)(R')(H) + X⁻]

sp³ hybridized carbon centre — transition state (pseudo pentavalent carbon) — sp³ hybridized carbon centre; inversion of configuration

Figure 3.6 The S_N2 reaction.

Worked example 3.3C

Provide a mechanism for the following transformation.

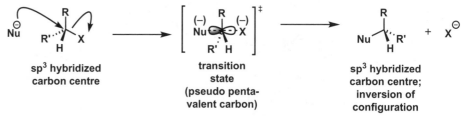

3.3 S$_N$1 AND S$_N$2

Solution

The carbon that connects to the leaving group is primary, therefore the mechanism must be S$_N$2. If it were S$_N$1, an unstable primary carbocation would be generated. The mechanism involves attack of the nucleophile onto the C to which the leaving group is bonded. The nucleophile (Cl$^-$) makes the new C—Cl bond at the same time as the C—O bond breaks; the process is concerted. More specifically, the incoming nucleophile attacks the C—O σ* orbital, which leads to cleavage of the corresponding C—O σ bond. It is due to the concerted nature of this reaction that in the transition state both the incoming nucleophile and leaving group are drawn carrying a partial negative charge.

→ For a refresher on chirality, see Chapter 2, section 2.5.

→ For a refresher on bonding and anti-bonding orbitals, see Chapter 1, section 1.3.

Worked example 3.3D

Provide a mechanism for the following transformation, assuming the reaction undergoes an S$_N$2-type reaction pathway. Suggest the stereochemical configuration of the product.

Solution

We are going to go through this example in two parts; in the first part we will look at the mechanism and in the second part we will look at the stereochemical implications of the mechanism. It is often useful to break up challenging questions into smaller steps in order to avoid confusion. In cases like this, trying to imagine the 3D positioning of groups whilst working out a mechanism could be tough.

The first thing to notice here is that the carbon centre at which the substitution occurs is secondary; this means that it can undergo either S$_N$1 or S$_N$2 substitution pathways. However, in the question we have been told that it reacts via an S$_N$2 mechanism therefore that is all we consider. In an S$_N$2 mechanism, the leaving group is displaced by the incoming nucleophile in a synchronous process. The mechanism is the same as in the previous example; the nucleophile attacks the C—Br σ* orbital, generating the transition state shown, where the incoming nucleophile and leaving group both have a partial negative charge. Finally, the leaving group is fully displaced and the product is produced.

Now that we have ascertained the reaction pathway, we can start to consider the stereochemical configuration of the product. In an S_N2 reaction, inversion of the original stereocentre is observed. This is due to the concerted nature of the reaction and that only one reaction pathway can be followed due to orbital constraints. The incoming nucleophile attacks the C—Br σ* orbital generating the transition state shown. In this case, the methyl group in the starting material was at the back, therefore in the transition state the methyl group remains at the back as the nucleophile has attacked in a trajectory between the methyl and hydrogen. Once the bromide has left, the resulting product has been inverted, rather like an umbrella being blown inside out. The final task is to determine the absolute configuration of the starting material and product. In this case the starting material has (R) configuration and the product (S).

→ Be careful—even if the starting material has an (R)-configuration, the product will not necessarily be of an (S)-configuration. You need to invoke the Cahn–Ingold–Prelog rules—see Chapter 2, section 2.3 for more information.

Question 3.8

Provide a mechanism for the following transformation.

Question 3.9

Provide a mechanism for the following transformation, assuming the reaction follows an S_N2 pathway. Suggest the stereochemical configuration of the carbon labelled with an asterisk.

▶ **Hint** Draw the mechanism first and then consider the stereochemical implications.

3.4 The impact of pK_a on leaving group ability

What is pK_a?

pK_a is the strength of an acid and is a measure of to what extent an acid is dissociated (Figure 3.7). In theory, all compounds that contain a C—H σ bond can act as an acid by donating that proton to a suitable base. There are many factors that can affect the acidity of an organic compound; the strength of the HA bond, the electronegativity of A, the stability of A⁻ compared to HA, and the solvent that the compound is in. When stating a pK_a value, the solvent in which the value was calculated is often given. Usually the solvent is water or DMSO (dimethylsulfoxide). pK_a values generally lie between −10 and 50, although there are some compounds with pK_a values that lie outside this range. If a compound has a low pK_a value it is a strong acid and the proton can be removed easily. If the pK_a value is high, however, the proton cannot be removed easily and its conjugate base is therefore strong.

3.4 THE IMPACT OF pK_A ON LEAVING GROUP ABILITY

Figure 3.7 An acid/base equilibrium.

$$HA + B \rightleftharpoons A^{\ominus} + {}^{\oplus}BH$$

We can determine if a proton can be removed easily by inspecting the C—H bond in question:

- A weaker H—A bond means the proton can be removed more easily so the pK_a is lower. This is not usually a limiting factor in pK_a values.
- Placing the negative charge of the anion on, or near, an electronegative atom e.g. oxygen is also stabilizing. In the case of ethanol, an alcohol, removal of the alcohol proton leads to an alkoxide which is stabilized by the electronegative oxygen. However, in the case of ethane, removal of a proton leads to a carbon-based anion (a carbanion) which is extremely unstable.
- The more stable the conjugate base, the more acidic the proton. Conjugate bases can be stabilized by a variety of means. If an anion can be distributed over a greater number of atoms it is more stable because each atom 'sees' less of the negative charge, leading to a lower pK_a value. This is known as resonance.
- Finally, hybridization is another important factor. The more s character an orbital has, the greater the degree of stability offered to the anion because of the spherical shape of the s orbital.

It should be noted that compounds with a low pK_a are excellent acids but very poor bases. The relationship between anion stability and pK_a in some selected compounds is shown in Figure 3.8.

→ For more information about resonance, see Chapter 1, section 1.8.

→ For a further explanation of this reaction, see Clayden et al. (2012).

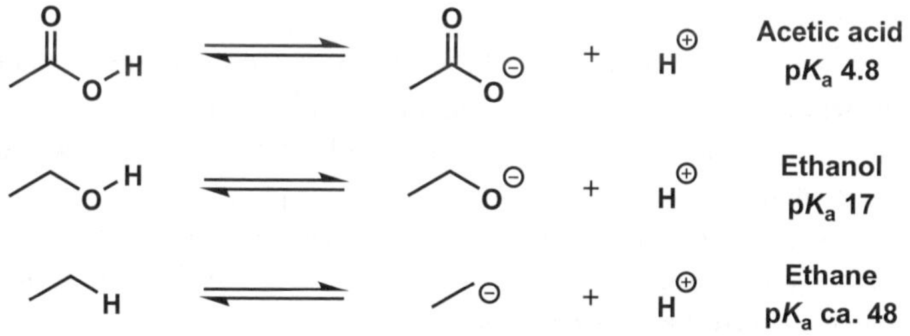

Acetic acid pK_a 4.8

Ethanol pK_a 17

Ethane pK_a ca. 48

Figure 3.8 A comparison of pK_a and anion stability for some selected compounds.

→ When discussing pK_a values of amines you need to take care because amines are amphoteric (can act as acids or bases). The pK_{aH} value of an amine refers to the **parent acid** (the protonated species) e.g. with ammonia, NH_3, pK_{aH} refers to NH_4^+. However, in ammonia the N—H bond can be broken to generate NH_2^-. To avoid confusion, if the structure is not drawn we will use the term pK_{aH}, which simply means the pK_a of the conjugate acid. For example with ammonia, the pK_{aH} is 9 (NH_4^+) and the pK_a is 33 (NH_3).

How is pK_a related to leaving group ability?

Generally speaking, if the pK_a of the conjugate acid of a leaving group is low, then it is a better leaving group. A list of approximate pK_a values is given in Table 3.1. The asterisk denotes the acidic proton.

Table 3.1 Useful approximate pK_a values.

	pK_a		pK_a		pK_a		pK_a
H*Cl	−7	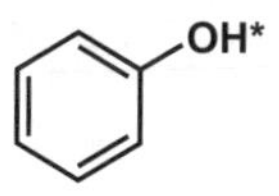 PhOH*	12	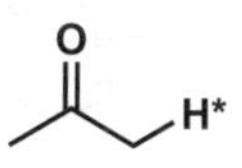 acetone (CH₃COCH₂H*)	19	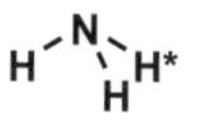 hydrazine (H₂N-NH*)	33
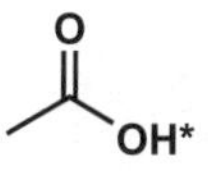 CH₃COOH*	5	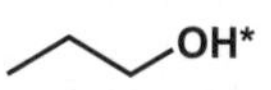 CH₃CH₂CH₂OH*	17	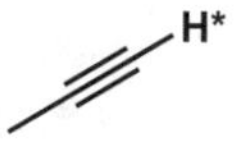 HC≡CH*	25	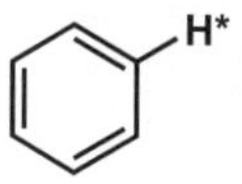 C₆H₅-H* (benzene)	42
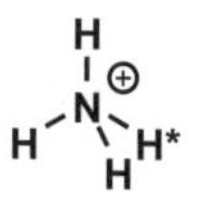 H₃N⁺H*	9	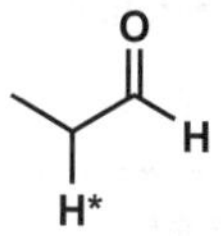 propanal (CH₃CH(H*)CHO)	18	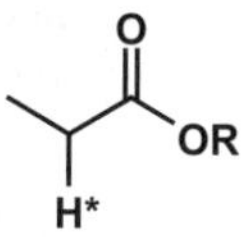 ester α-H (CH₃CH(H*)COOR)	25	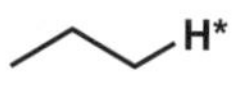 CH₃CH₂CH₂H*	48

Worked example 3.4A

Compare the acidity of the following protons marked with an asterisk. Suggest reasons for their differing acidities.

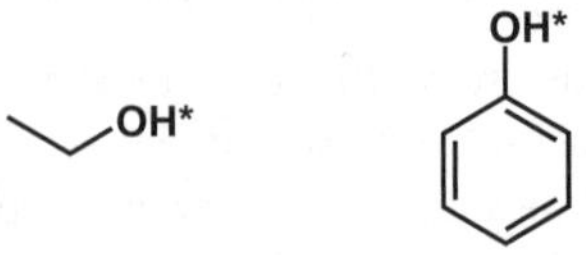

Ethanol (CH₃CH₂OH*)   **Phenol** (C₆H₅OH*)

Solution

In this example both compounds contain an alcohol functional group. In the first example, ethanol, removal of the proton generates a negative charge on oxygen. This leads to some stabilization due to the electronegativity of oxygen therefore the proton is reasonably acidic; however, the negative charge is completely localized on the oxygen atom.

Ethanol
pK_a 17

> For a refresher on resonance stabilization, see Chapter 1, section 1.8.

In the second example, phenol, removal of the proton generates a negative charge on oxygen which can be distributed into the benzene ring due to resonance stabilization. This is an extremely effective way of stabilizing a negative charge and is reflected in the pK_a values; the pK_a of ethanol is approximately 17 and that of a phenol is 12, i.e. phenol is more acidic and will lose its proton more readily as the conjugate base is more stable. Additionally, more stabilization is offered because the carbon atoms in phenol are sp^2, therefore they have increased s character so are better able to stabilize a negative charge.

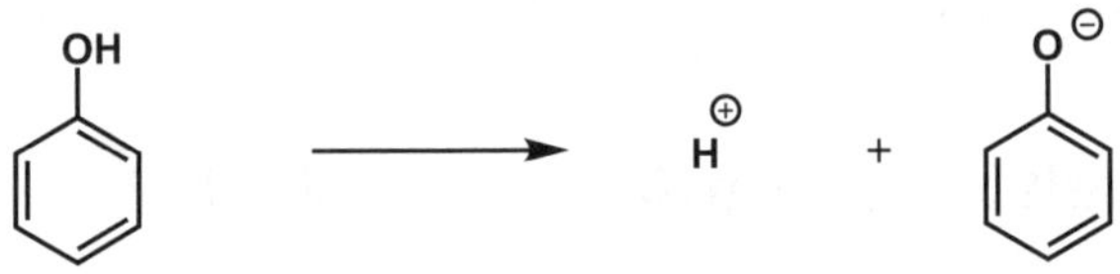

Phenol
pK_a 12

Overall the series is as follows:

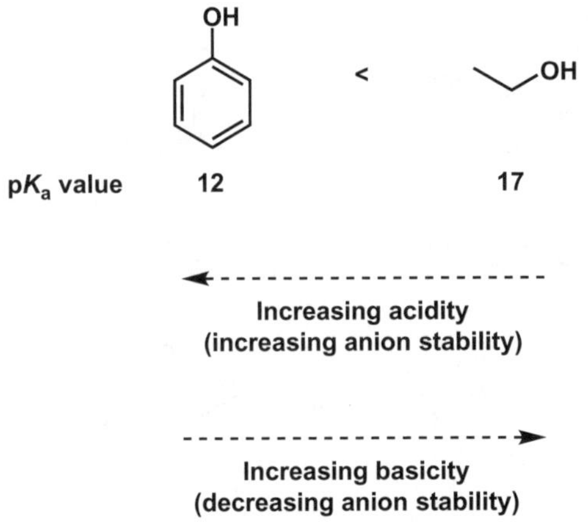

Phenol < Ethanol

pK_a value   12   17

← Increasing acidity (increasing anion stability)

→ Increasing basicity (decreasing anion stability)

3.4 THE IMPACT OF pK_A ON LEAVING GROUP ABILITY

Worked example 3.4B

Suggest which of the following molecules has the highest pK_a, i.e. is the least acidic.

Solution

In this case we have a proton in three different hybridization states: sp^3, sp^2, and sp. When we remove the proton indicated we will generate the corresponding sp^3, sp^2, or sp anion. These have different relative stabilities due to the amount of s character present in the orbital: as s character increases, the anion is held closer to the nucleus so the electrons are more tightly held and are therefore more stabilized, resulting in a lower energy species. This is reflected in the pK_a. As s character increases, i.e. as we travel from sp^3 to sp^2 to sp, anion stability increases and pK_a decreases.

→ In chemistry, generally speaking, the lower in energy something is, the more stable it is.

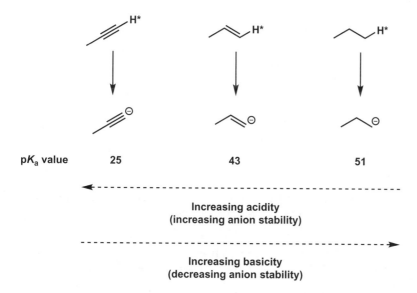

Worked example 3.4C

Taking into account the stability of the anion, suggest which of these leaving groups will be the best.

Solution

The pK_a of the alkoxide is approximately 17, the sulfonate −3, and the *p*-nitro phenol 8. The compound with the lowest pK_a is the best leaving group and is more able to stabilize a negative charge, therefore the sulfonate is the best leaving group and the alcohol the worst. This is utilized within organic chemistry, as a sulfonate is often used as a leaving group in substitution and elimination reactions.

If we did not know the pK_a values for each, it is possible to compare the compounds in order to predict which would be the better leaving group. In all examples the anion is on oxygen so it is stabilized by the electronegativity of oxygen, therefore we need to inspect each molecule more closely to see if there are likely to be any differences in stability, and therefore pK_a.

If we compare the sulfonate and the phenoxide with the ethoxide, respectively, the anion can undergo resonance stabilization in the phenoxide and sulfonate whereas it cannot in the ethoxide. This means that these anions are both more stable than the ethoxide. Comparing the phenoxide and sulfonate, the anion in the sulfonate is strongly stabilized by the electron-withdrawing SO_2 group. In the phenoxide the anion can resonate around the benzene ring into the nitro group, but this is not as stabilizing because the nitro group is further away so the electron-withdrawing effect is weaker.

❓ Question 3.10

Why is there such a difference in pK_{aH} values of the following: butylamine (pK_{aH} 10.7); dibutylamine (pK_{aH} 11.3); and tributylamine (pK_{aH} 9.9)?

Butylamine   Dibutylamine   Tributylamine

▶ **Hint** Consider hydrogen-bonding with the solvent and the inductive effect of the alkyl chains.

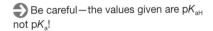
Be careful—the values given are pK_{aH} not pK_a!

❓ Question 3.11

Which of the following compounds will have the lower pK_a?

3.5 Synoptic questions

> **Question 3.12**
>
> What is the most likely mechanistic pathway for the following reaction? Suggest the stereochemical configuration of the product.
>
>

> **Question 3.13**
>
> What is the product of the following reaction? Draw the curved arrow mechanism and determine the stereochemical configuration of the product.

> **Question 3.14**
>
> Consider the following reaction. What would happen to the rate if:
> (a) the substrate concentration was doubled?
> (b) the nucleophile concentration was halved?
>
>

> **Question 3.15**
>
> What products could be made under the reaction conditions shown?

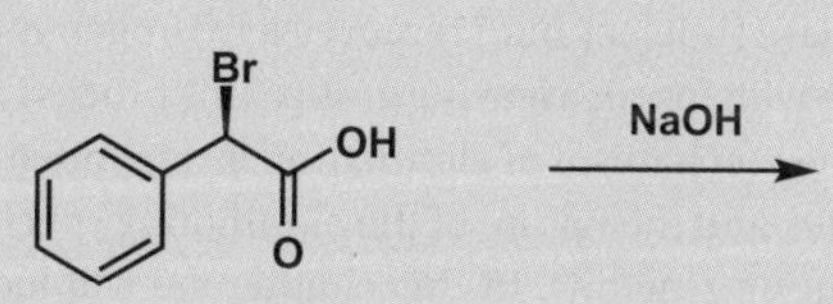

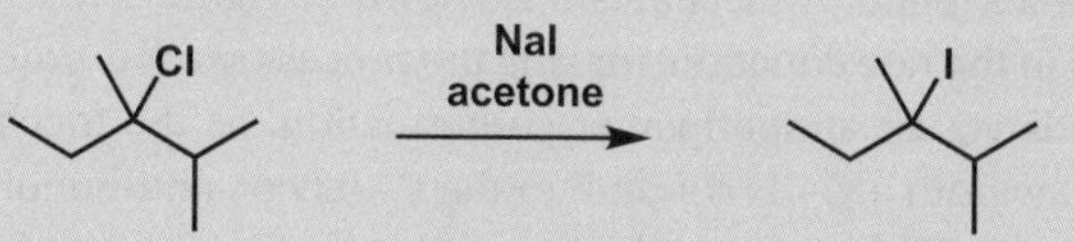

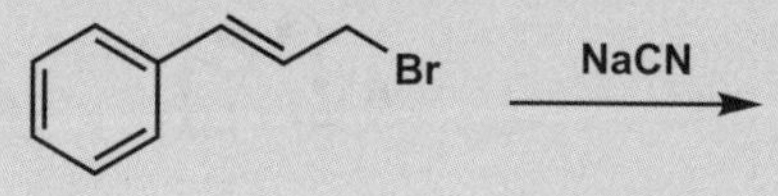

References

Burrows, A., Holman, J., Parsons, A., Pilling, G., and Price, G. (2013) *Chemistry³*, 2nd edn (Oxford University Press, Oxford).

Clayden, J., Greeves, N., and Warren, S. (2012) *Organic Chemistry*, 2nd edn (Oxford University Press, Oxford).

4
Elimination reactions

4.1 Synthesis of alkene via elimination (E2, E1, and E1cB)

In elimination reactions, a base (nucleophile) removes a hydrogen nucleus (proton), resulting in the formation of an alkene with loss of a leaving group. This process can occur by three mechanisms, E2, E1, and E1cB. The mechanism of elimination is dependent on factors including the solvent, nature of nucleophile, and the nature of the eliminating species. Elimination reactions often compete with substitution reactons. In this chapter we will look at the major products of elimination reactions, but be aware that in practice some of these experiments would yield several impurities, including substitution products.

E2 eliminations

E2 elimination is so-named because it is a bimolecular elimination, involving a concerted mechanism where a base removes a proton, with concomitant displacement of a leaving group and the formation of a π-bond. A mechanism is shown in Figure 4.1. Both the base and the reagent are included in the rate equation, making the process second-order, like S_N2 reactions. The reaction proceeds via an antiperiplanar conformation, as the mechanism requires the movement of electrons from a C—H σ-bond, to the C—LG σ*-anti-bonding orbital, resulting in the breakage of the C—LG bond, and the formation of a C—C π-bond. The E2 mechanism is favoured if a strong base (e.g. an alkoxide or lithium diisopropylamide (LDA)) is used with reagents where the leaving group sits in a primary or secondary position.

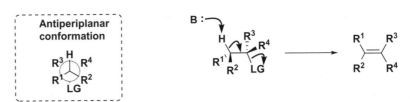

Figure 4.1 A general mechanism for an E2 elimination. The reagent takes an antiperiplanar conformation in order to allow the movement of electrons into the σ* anti-bonding orbital, yielding a stereospecific product. A strong base is typically used.

E1 eliminations

E1 eliminations are unimolecular, with only the concentration of the molecule undergoing elimination contributing to the rate equation. E1 eliminations, like S_N1 reactions, involve two steps: the leaving group leaves (rate determining), then the proton is removed, forming an alkene (Figure 4.2). Unlike E2 reactions this process does not require an antiperiplanar conformation, however steric effects can lead to these reactions being stereoselective; usually *E*

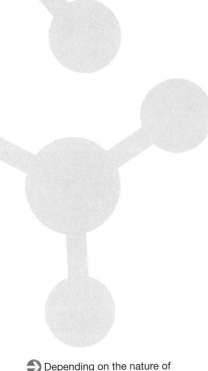

➔ Depending on the nature of the nucleophile, either elimination or substitution can occur. Bulky nucleophiles, strong bases, and high temperatures favour elimination over substitution, as does increasing the concentration of base.

➔ E2 eliminations prefer to yield a more substituted alkene (Zaitsev product), as they are more stable, but steric hindrance when using bulky bases can force the process to yield a less substituted alkene (Hofmann product).

➔ The diagram showing the antiperiplanar conformation in Figure 4.1 is called a Newman projection. This type of projection demonstrates the relative position of substituents on two adjacent atoms, and is drawn as if you are looking down the length of the bond.

➔ E1 eliminations are regioselective for the more substituted alkene (Zaitsev product).

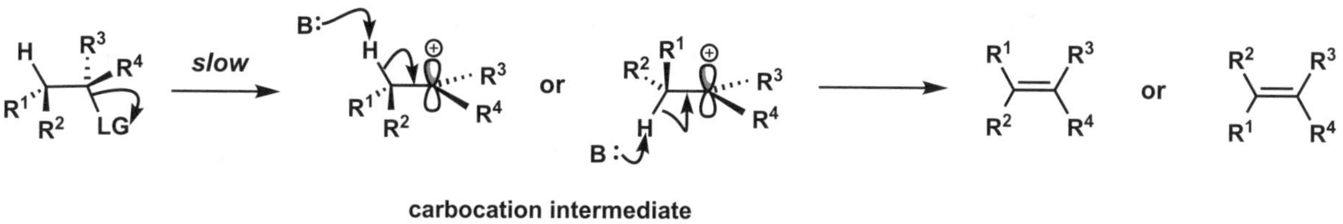

4.1 SYNTHESIS OF ALKENE VIA ELIMINATION (E2, E1, AND E1cB)

Figure 4.2 The mechanism of an E1 elimination. The carbocation intermediate is planar and sp² hybridized, allowing elimination from either side of the vacant p-orbital, which potentially yields both the *E* and *Z* products. During the elimination step, the C—H σ-bond must be aligned with the vacant p-orbital of the cation.

alkenes are favoured due to this effect. E1 reactions typically occur when the structure of a reagent means that the carbocation formed sits on a tertiary or secondary position, and where the carbocation intermediate may be stabilized by surrounding substituents. E1 reactions require a good leaving group and a weaker base (e.g. water or an alcohol), or no base at all!

E1cB eliminations

The E1cB mechanism is a two-step process that proceeds with the formation of a carbanion, and subsequent displacement of a leaving group (Figure 4.3). The reactions are first order with respect to the conjugate base. E1cB occurs when a a molecule with a poor leaving group is treated with a strong base, and is only possible when the carbanion formed is stabilized by an electron-withdrawing group (commonly a carbonyl). As in E1 reactions, the E1cB elimination is regioselective based on steric effects, often yielding a major *E* product.

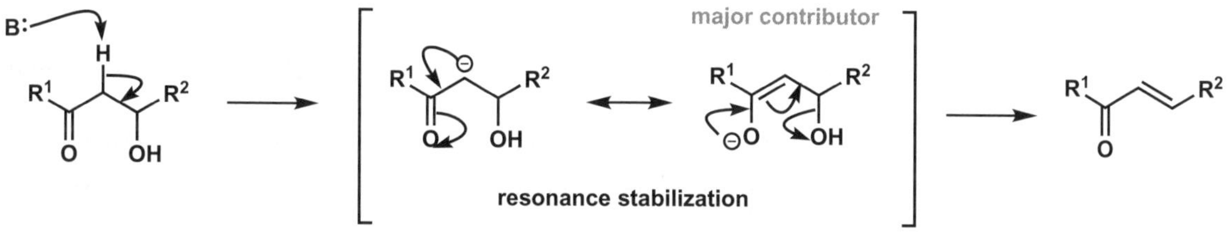

Figure 4.3 The mechanism for an E1cB elimination. The carbanion intermediate is stabilized by an electron withdrawing group—in this case, a carbonyl.

Worked example 4.1A

The reaction below yields a single product through an elimination reaction. What is the product, and by which mechanism is it formed?

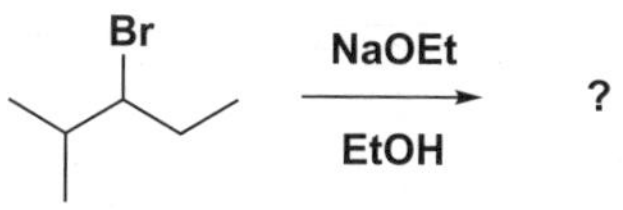

> The alkoxide and corresponding alcohol conditions used for this reaction are very common conditions for elimination reactions. Please note that if the alkoxide and leaving group are not sterically hindered, then S_N2 reactions may occur.

Solution

If we look at the reaction conditions, then we can note two things. First, a pretty strong alkoxide base is used, which would favour the E2 mechanism, and second the leaving group (Br) is in a secondary position. This means that were the reaction to proceed via an E1 mechanism, the carbocation would be secondary, which is not very stable in the absence of electron donating groups. These point to the likely mechanism being **E2**.

Now, when trying to decide the likely product, we need to identify which protons could be eliminated. These must be adjacent to the leaving group, giving us three possible protons to choose from.

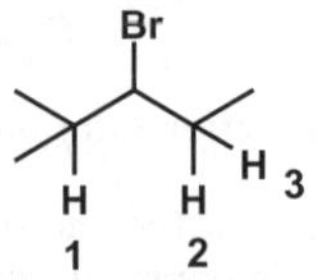

Eliminating proton **1** would lead to a trisubstituted alkene, whilst eliminating **2** or **3** would give a disubstituted alkene. Zaitsev's rule states that alkenes of higher substitution are more stable, so eliminating proton **1** is preferable. Therefore the product would be:

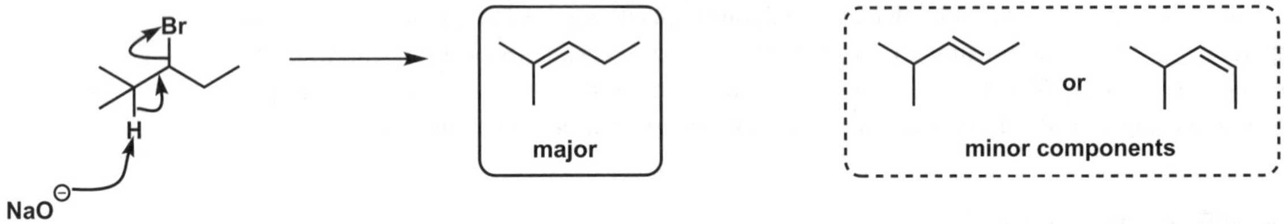

Worked example 4.1B

Heating 3-iodo-3-phenylpropanoic acid in water yields two products, (*E*)-3-phenylprop-2-enoic acid (cinnamic acid) and (*Z*)-3-phenylprop-2-enoic acid.

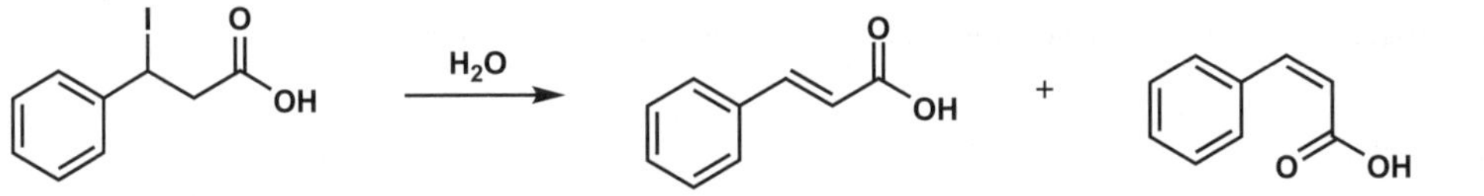

(a) Provide a mechanism for this reaction, explaining why two stereoisomers are formed.
(b) This reaction is stereoselective. Which do you expect to be the major product, and why?

Solution

(a) In 3-iodo-3-phenylpropanoic acid, iodine is a good leaving group and is located in a position that yields a carbocation which can be stabilized by resonance from the phenyl group. The only reagent is water, which is a polar protic solvent. These conditions favour the E1 elimination mechanism. This reaction mechanism will proceed with formation of a carbocation by the iodine atom leaving as iodide (I⁻). This carbocation is free to rotate around the C—C bond, so the electrons in either of the two adjacent C—H bonds to be eliminated can align with either lobe of the vacant p-orbital of the carbocation. This yields two potential products when the proton is abstracted during the elimination step.

→ Note that even if there was only one abstractable proton adjacent the carbocation we would still synthesize two products, as the C—H can align with either lobe of the carbocation's p-orbital.

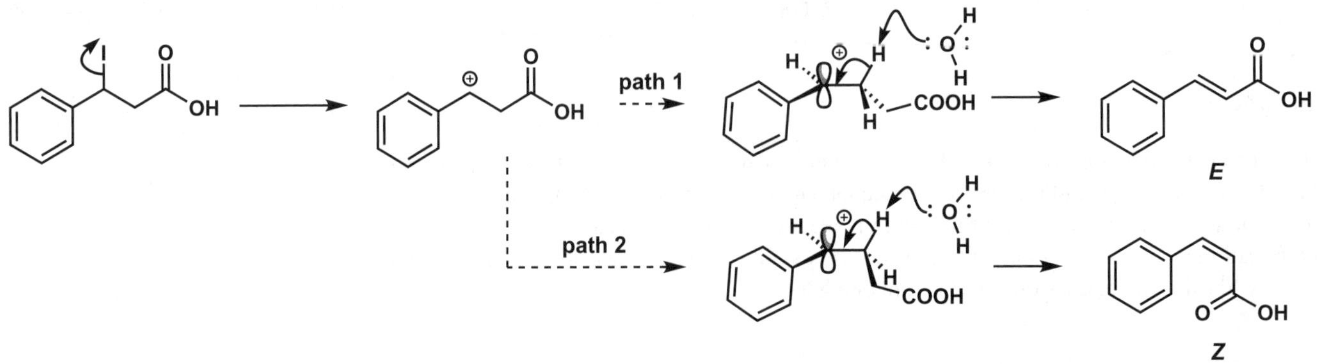

(b) E1 reactions are stereoselective due to steric interaction between adjacent substituents on the carbocation intermediate. If we draw Newman projections of the two intermediates we can make a judgement about which is less sterically favourable.

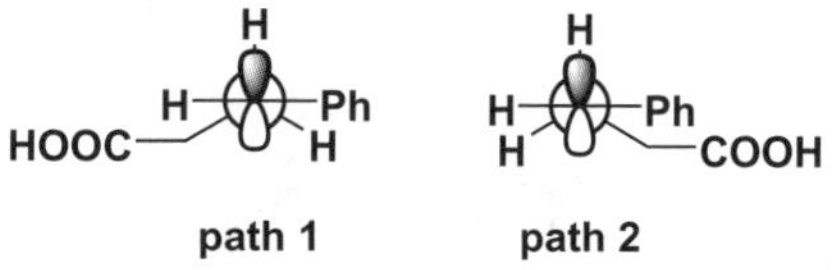

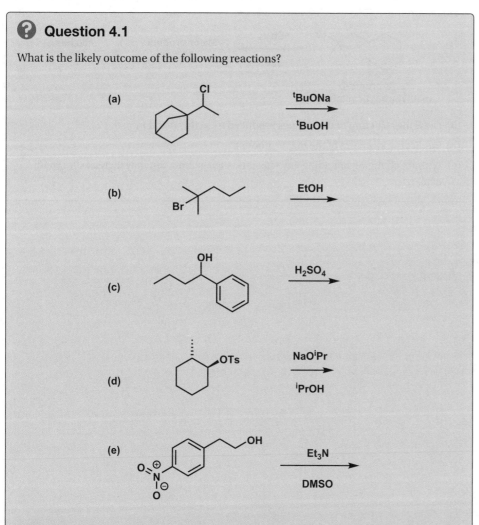

The major bulky groups in these projections are the phenyl and carboxylic acid groups. This means that there is more steric hindrance in path 2, which aligns the groups with each other. Thus, path 1 is more favourable, making the *E* stereoisomer, cinnamic acid, the major product.

Question 4.1

What is the likely outcome of the following reactions?

(a) [structure with Cl] → ^{t}BuONa / ^{t}BuOH

(b) [structure with Br] → EtOH

(c) [structure with OH] → H_2SO_4

(d) [cyclohexane with OTs] → NaO^{i}Pr / ^{i}PrOH

(e) [structure with OH and nitro group] → Et$_3$N / DMSO

▶ **Hint** When approaching these questions, try to look at the choice of base and the nature of the leaving group. A strong base will often give E2 reactions, whilst a good leaving group that results in the formation of a relatively stable carbocation will favour E1.

> **Question 4.2**
>
> How many different alkenes could be formed from the E2 elimination of the compound below?

> **Question 4.3**
>
> In the laboratory, you stir a solution of (2-bromo-2-methylpropyl)cyclohexane in ethanol at room temperature for 24 h. This yields a minor fraction (25%) of alkene products, but also a major product (75%) which is not an alkene. 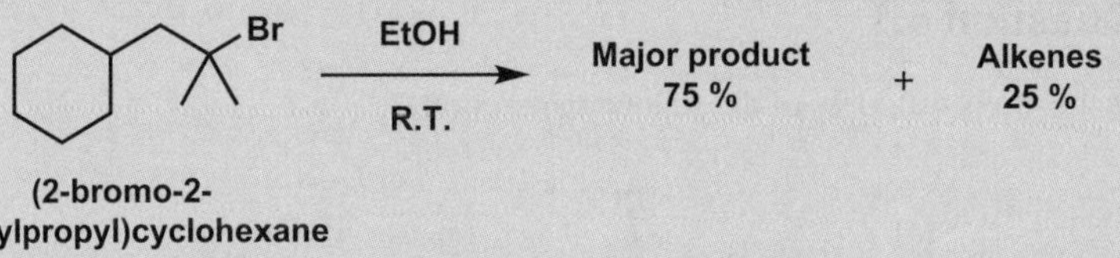
>
> (2-bromo-2-methylpropyl)cyclohexane + EtOH, R.T. → Major product 75 % + Alkenes 25 %
>
> (a) Draw the structure of the alkenes produced from this reaction.
> (b) Predict the structure of the major product.
> (c) Without changing or adding any reagents, suggest how you may increase the yield of alkenes.

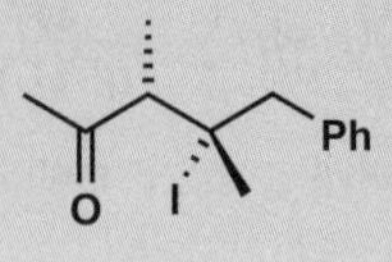

5
Reactions of unsaturated compounds

5.1 Electrophilic addition

In general, electrophilic addition reactions occur between an electrophile and an alkene, though other groups possessing π-bonds may also react in this way. Electrophilic addition results in the π-bond breaking with the formation of two new σ-bonds, with the electrophile 'adding' to the alkene to form a new, larger, molecule. An example electrophilic addition is shown in Figure 5.1, in which hydrogen bromide 'adds' to an alkene π-bond. Common reagents which undergo electrophilic addition to alkenes are hydrogen halides, alcohols, and water. Acid catalysts are often used which act as an electrophile, reacting with the alkene in the first instance to form a carbocation, which is then attacked by a nucleophile.

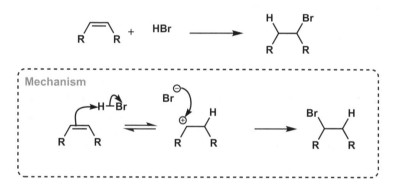

Figure 5.1 An alkene undergoes electrophilic addition when treated with hydrogen bromide.

Regioselectivity in electrophilic addition

As the carbocation formed during electrophilic addition could reside at either carbon of the π-bond, electrophilic additions are regioselective, giving a major product which has the more stable carbocation in its transition state. For instance, electrophilic addition to propene gives a

➔ Markovnikov's rule is used to predict the regioselectivity of electrophilic additions. In essence it states that when the general electrophile 'HX' undergoes addition to an alkene, then the proton is added to the carbon atom with the most hydrogen atoms. While this is often the case, it is not always true. Products which adhere to this rule are often called the 'Markovnikov' product, whilst those which don't are called 'anti-Markovnikov'. Anti-Markovnikov products are often seen in radical-mediated reactions.

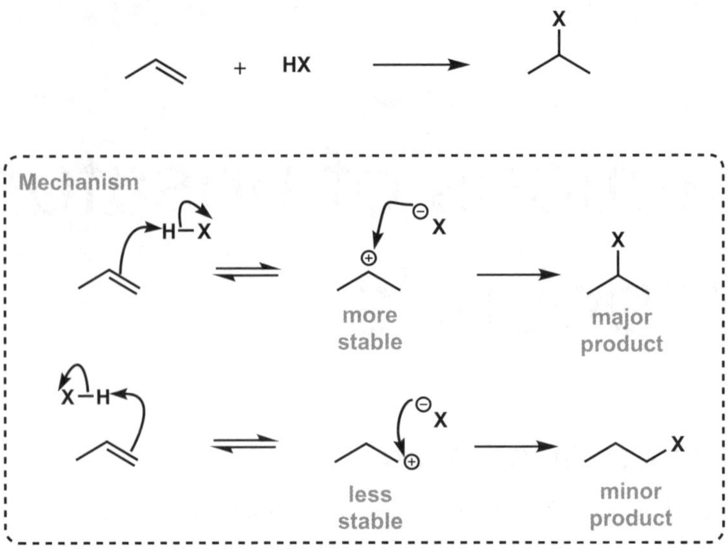

5 REACTIONS OF UNSATURATED COMPOUNDS

Figure 5.2 The stability of the carbocation intermediate dictates the regioselectivity of electrophilic addition.

major product which goes through the secondary carbocation transition state, which is more stable than the primary (Figure 5.2).

Halogen addition reactions

> The term 'vicinal' means that the functional groups are bonded to carbon atoms adjacent to each other. 'Geminal' means that they are bonded to the same carbon atom.

Electrophilic addition reactions occur between alkenes and diatomic halogen molecules (Br_2, Cl_2, or I_2), to form a vicinal dihalide. In this reaction, the electron density of the alkene π-bond induces a dipole on the halogen molecule, making it electrophilic. The reaction mechanism is shown in Figure 5.3. Movement of electrons from the π-bond into the σ* anti-bonding orbital results in the σ-bond breaking, and the formation of a new σ-bond between the halogen and both carbon atoms, and a negatively-charged halide ion. The positively charged intermediate is known as a halonium ion, and is attacked in the manner of a nucleophilic addition to yield a product with two new vicinal halide groups.

> A non-nucleophilic solvent, like dichloromethane, must be used for this reaction. If this reaction is conducted in the presence of water, it can attack the halonium intermediate, giving a 'halohydrin', which has a halogen and hydroxyl group vicinal to each other.

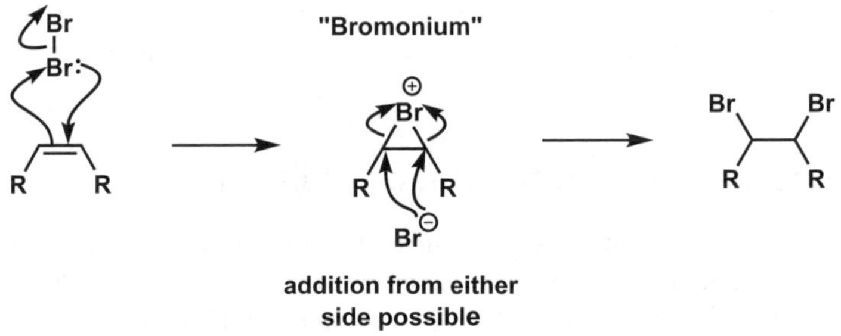

Figure 5.3 Treatment of an alkene with bromine results in a dibrominated product via electrophilic addition.

Worked example 5.1A

What is the major product for the following reaction?

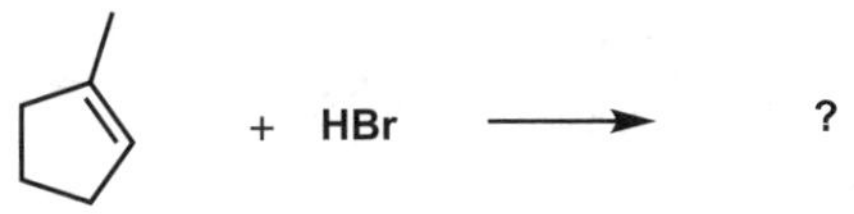

Solution

In this chapter we learn that hydrogen halides undergo electrophilic addition to alkenes, of which this question is an example. The first step of this reaction involves the attack of the nucleophilic alkene to the electrophilic proton of the hydrogen bromide. This could result in carbocations I or II. I is more stable as it has the greater number of alkyl substituents, so the reaction will proceed primarily via this intermediate. The bromide ion formed from the initial step can then add to the carbocation formed to give 1-bromo-1-methylcyclopentane as the major product.

Worked example 5.1B

Two equivalents of styrene undergo electrophilic addition with ethylene glycol (ethan-1,2-diol) in the presence of sulfuric acid. What is the major product?

Solution

We know that this is an electrophilic addition reaction, so first of all we will identify the electrophile and nucleophile. The nucleophile must be electron-rich, and reactive, so styrene is a good candidate. Styrene is electron-rich at the aromatic phenyl ring and vinyl double bond. However, the aromaticity of the ring means that it will not undergo electrophilic addition at that point, which would destroy the aromatic π-system. Styrene is therefore nucleophilic at the vinyl double bond. The electrophile, in the first instance, will be a proton from the sulfuric acid catalyst, since ethylene glycol is a weaker acid. If we draw out the first addition step, we have two possible carbocation intermediates, one at the primary position, and one at the secondary position. As discussed previously, the secondary carbocation is more stable due to +I effects from the alkyl substituents, but also, in this case, the phenyl ring is able to add extra stabilization through resonance.

→ Aromatic molecules will undergo electrophilic aromatic substitution, rather than addition, but this requires additional reagents. See Chapter 6 for more information.

At this point, the nucleophilic ethylene glycol is able to add to the carbocation intermediate, followed by the loss of a proton to yield an ether group between the two reagents and restore the acid catalyst. This would lead to a racemic mixture of products, since the ethylene glycol is able to add to either lobe of the carbocation's p-orbital.

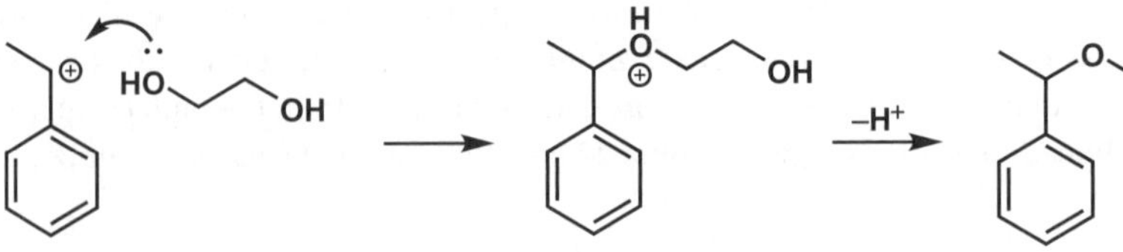

We must resist the temptation to finish here, because we've been told that two equivalents of styrene have been used, so there is still some reagent to react with the alcohol we've just formed. This will proceed as an electrophilic addition, just like last time. The major product drawn here has no stereochemistry defined, but would be a racemic mixture of the four possible diastereoisomers/enantiomers.

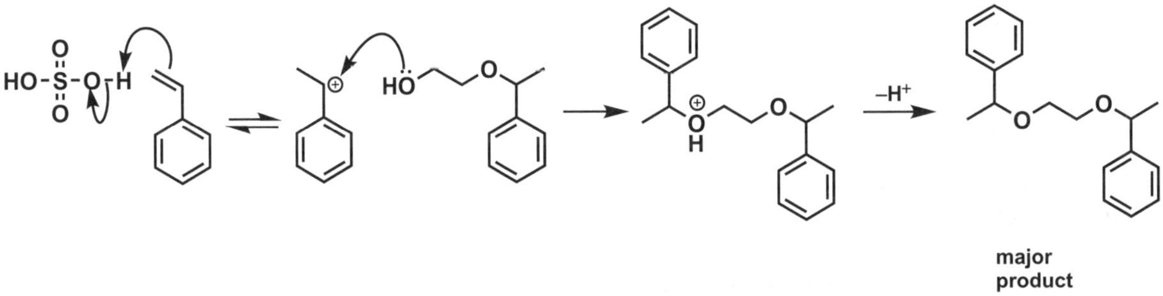

major product

Question 5.1

Predict the major product for the following reactions, assuming no side-reactions, and comment on its stereochemistry. Include a mechanism for each reaction.

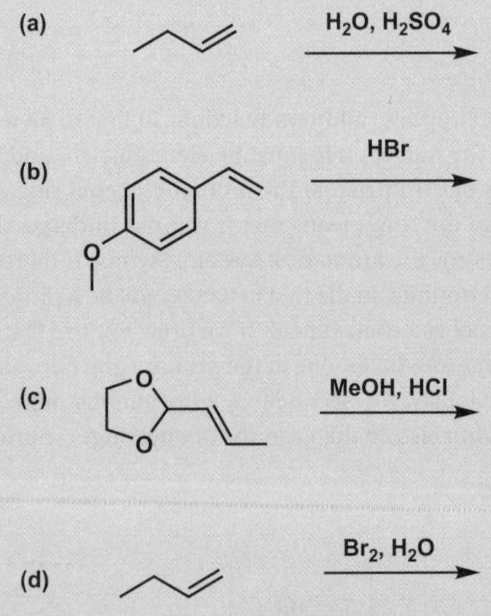

(a) propene + H_2O, H_2SO_4

(b) 4-methoxystyrene + HBr

(c) 2-(prop-1-en-1-yl)-1,3-dioxolane + MeOH, HCl

(d) propene + Br_2, H_2O

▶ **Hint** Remember to think about resonance and inductive effects on the carbocation intermediate that may be stabilizing or destabilizing.

▶ **Hint** Ensure that you are careful with stereochemistry—review your answers to ensure that you have included all possible stereoisomers in your answers. Each carbocation p-orbital can be attacked by a nucleophile from either lobe.

Question 5.2

Treating cyclohexene with Br_2 yields *trans*-1,2-dibromocyclohexane. Justify why the *cis* isomer is not formed.

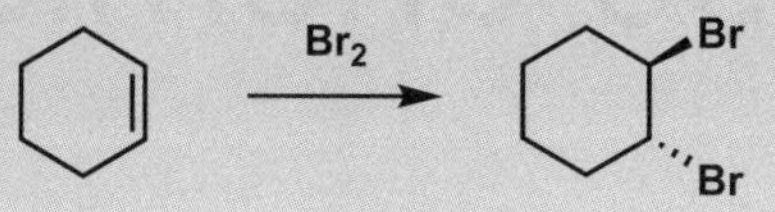

▶ **Hint** It will help to draw out the mechanism for this reaction.

6
Aromatic chemistry

6.1 Electrophilic aromatic substitution

What is electrophilic aromatic substitution?

Electrophilic aromatic substitution (S_EAr) is where a functional group on an aromatic ring, usually a proton, is replaced by an electrophile.

In order to undergo reaction with the electrophile, the aromatic ring must be electron-rich. This is because the mechanism requires the cloud of electrons on the benzene ring to attack the electrophile; if the ring were electron-deficient, it would not be able to do this as readily (Figure 6.1). The mechanism proceeds via a carbocation intermediate, also known as the Wheland intermediate. This intermediate can then lose a proton, restoring the aromaticity.

→ For more information about electrophiles, see Chapter 3, section 3.1.

Figure 6.1 Electrophilic aromatic substitution general scheme.

Worked example 6.1A

Provide a mechanism for the following reaction. Suggest why nitration only occurs once.

Solution

The overall reaction shows one of the protons on the benzene ring has been replaced by an NO_2 group. In order for this reaction to work, we need to generate a species that can act as a source of NO_2; we need to consider formation of an active electrophile. Reaction of nitric acid (HNO_3) with sulfuric acid (H_2SO_4) generates a nitronium ion (NO_2^-). In this sequence, nitric acid is protonated and a water molecule is subsequently eliminated.

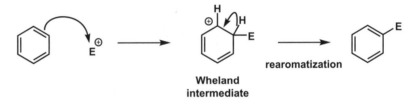

nitronium ion

With the electrophile prepared, the S$_E$Ar step can be considered. The pi-cloud of the benzene ring can attack the nitronium ion, a very strong electrophile, to generate the Wheland (carbocation) intermediate. The high-energy carbocation is no longer aromatic, so a proton leaves in order to regain aromaticity.

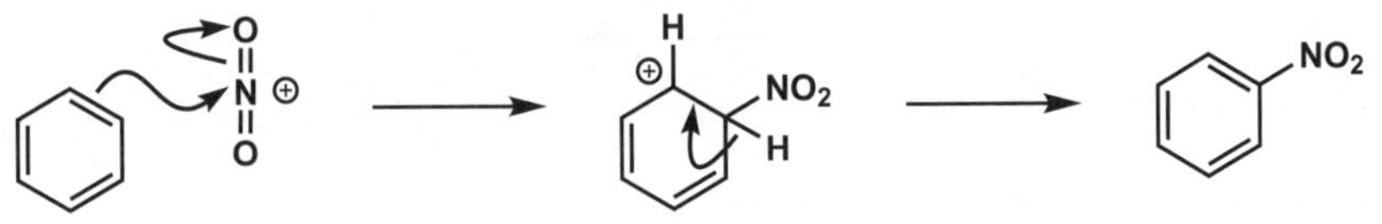

Finally, the reason that the ring is only nitrated once is because the nitro group is strongly electron withdrawing, therefore it deactivates the ring, preventing it from attacking other electrophiles. If you recall from earlier, S$_E$Ar requires an electron-rich substrate. It is possible to nitrate the ring again but it requires extremely forcing conditions: fuming HNO_3, concentrated H_2SO_4, and heating to 100 °C!

Worked example 6.1B

Provide a mechanism for the following reaction.

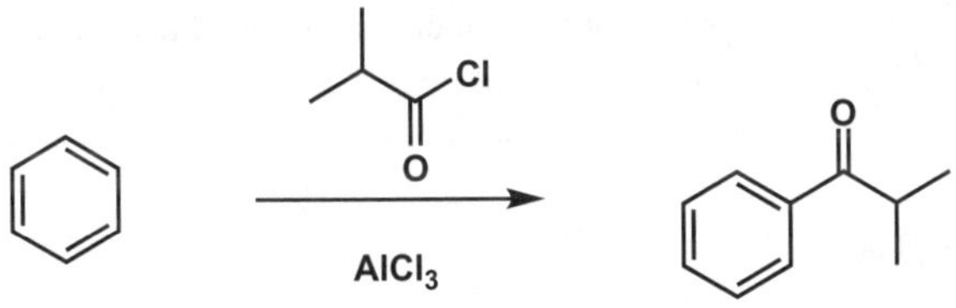

Solution

There are two aspects in this question that will help you to answer it: the first is that the group added to the benzene ring is almost the same as that over the arrow, but without the Cl. The second is that there is an acyl chloride and a Lewis acid ($AlCl_3$). These two pieces of information point to a Friedel–Crafts type mechanism.

The first step is coordination of the Lewis acidic aluminium species to the chloride, followed by a lone pair on the carbonyl oxygen eliminating $AlCl_4^-$, to generate a highly reactive acylium ion. Once we have generated the acylium ion, we are in a familiar S$_E$Ar mechanism; the ion is attacked by the benzene ring generating the Wheland carbocation intermediate before the ring undergoes rearomatization, providing the final product.

> For a refresher on Lewis acids, see Chapter 3, section 3.2.

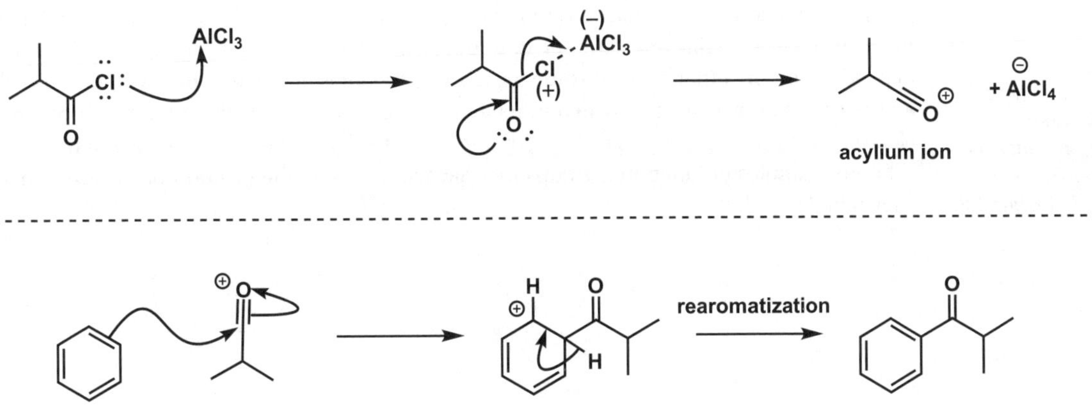

6 AROMATIC CHEMISTRY

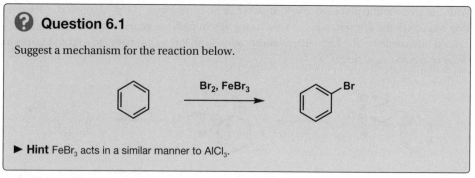

Question 6.1

Suggest a mechanism for the reaction below.

▶ **Hint** FeBr$_3$ acts in a similar manner to AlCl$_3$.

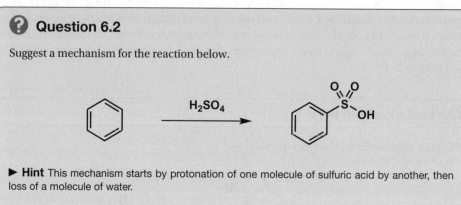

Question 6.2

Suggest a mechanism for the reaction below.

▶ **Hint** This mechanism starts by protonation of one molecule of sulfuric acid by another, then loss of a molecule of water.

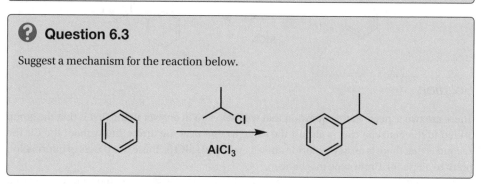

Question 6.3

Suggest a mechanism for the reaction below.

➔ For more information about directing and activating effects, see Clayden et al. (2012).

➔ Inductively means electron donation/withdrawal through σ bonds, and mesomerically means electron donation/withdrawal through π bonds. This is covered in Chapter 1, sections 1.6 and 1.8.

6.2 Effects of directing groups on S$_E$Ar

What is a directing group?

Once a benzene ring is substituted, different positions will be favoured to be reactive. A directing group is a functional group that can guide a reagent to react at a specific position on a molecule. In terms of electrophilic aromatic substitution, S$_E$Ar, the position that undergoes substitution depends upon the other ring substituents. If a substituent is electron donating, then the *ortho* and *para* positions will be activated in relation to the donating substituent. If a substituent is electron-withdrawing, the positions substituted are *meta* in relation to that substituent.

Some examples of functional groups that have an effect on the position of substitution are shown in Figure 6.2.

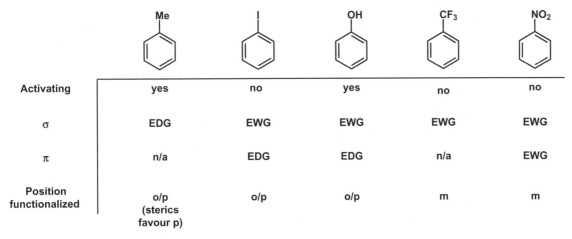

Figure 6.2 The directing effects of some commonly encountered substituents. EWG means 'electron withdrawing group' and EDG means 'electron donating group'.

Worked example 6.2A

Suggest where the following molecules will undergo electrophilic aromatic substitution.

→ The nitro group is drawn as shown.

It is important that you learn to draw it out like this as it is commonly drawn incorrectly in exams!

Solution

If we consider the first example, nitrobenzene, the nitro group is strongly deactivating in both a σ and π manner (inductively and mesomerically respectively), because it pulls electron density towards itself. There are two implications of this deactivating effect: first, there is less electron density on the ring at the *ortho* and *para* positions (with respect to the nitro group) due to the mesomeric (π) electron withdrawing nature of the nitro group, therefore any electrophile will preferentially add at the *meta* position. Second, when considering any resonance forms of the Wheland intermediate, we need to look at possible locations of the positive charge.

If we follow Pathway A, where the electrophile adds *para* with respect to the nitro group, there is a resonance form where the positive charge can be located on the carbon directly adjacent to the nitro group; an extremely unfavourable interaction. This is also true if the electrophile adds *ortho* to the nitro group. However, if the electrophile adds *meta* to the nitro group (Pathway B), the positive charge generated cannot be located directly adjacent to the nitro group and the intermediate generated is therefore much more stable. Both of these effects (the electron withdrawing nature of the nitro group and the more stable carbocation intermediate) lead to preferential formation of the *meta* product.

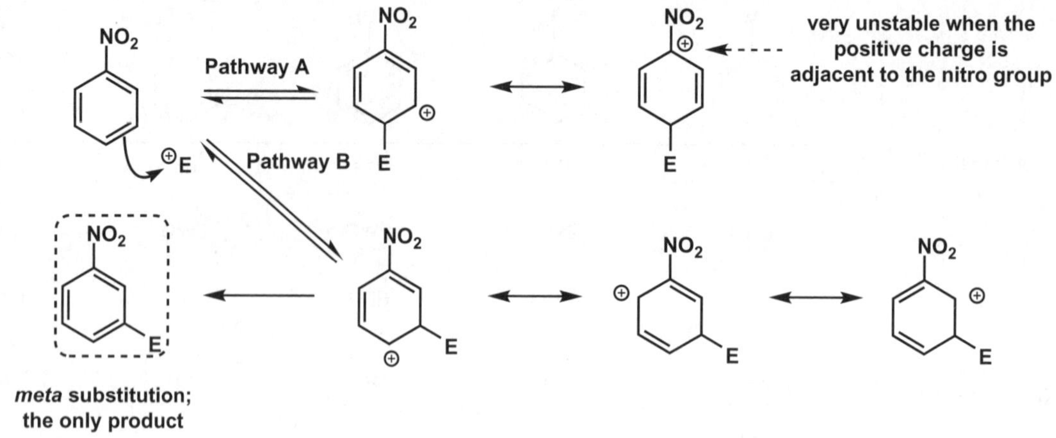

The second example we need to consider is phenol. In phenol, the OH is electron withdrawing through the σ bonds due to the electronegativity of oxygen (inductive effect), but electron donating through the π bonds (mesomeric effect). However, the dominant effect is the π electron donation, therefore the OH group can be considered an activating group. As with the first example, there are two things to consider. First, the electron-inducing nature of the OH group means that the *ortho* and *para* positions are electron-rich compared to the *meta* position so these positions are likely to attack an electrophile.

Second, we need to consider the intermediate carbocation species. Both *meta* or *para* attack onto an electrophile look plausible as resonance allows the positive charge to be moved around the ring, which is stabilizing. We therefore need to look into these possibilities in more detail. If we follow Pathway A, where the electrophile attacks *meta* to the OH, the positive charge is never located adjacent an oxygen which we may consider to be a more favourable interaction. However, if we follow Pathway B, where the positive charge can be adjacent the oxygen, one of the oxygen lone-pairs can be used to stabilize the positive charge by π resonance, as shown. This is also true if the phenol attacks in the *ortho* position. Overall, addition of an electrophile to phenol will either be *ortho* or *para* to the OH group.

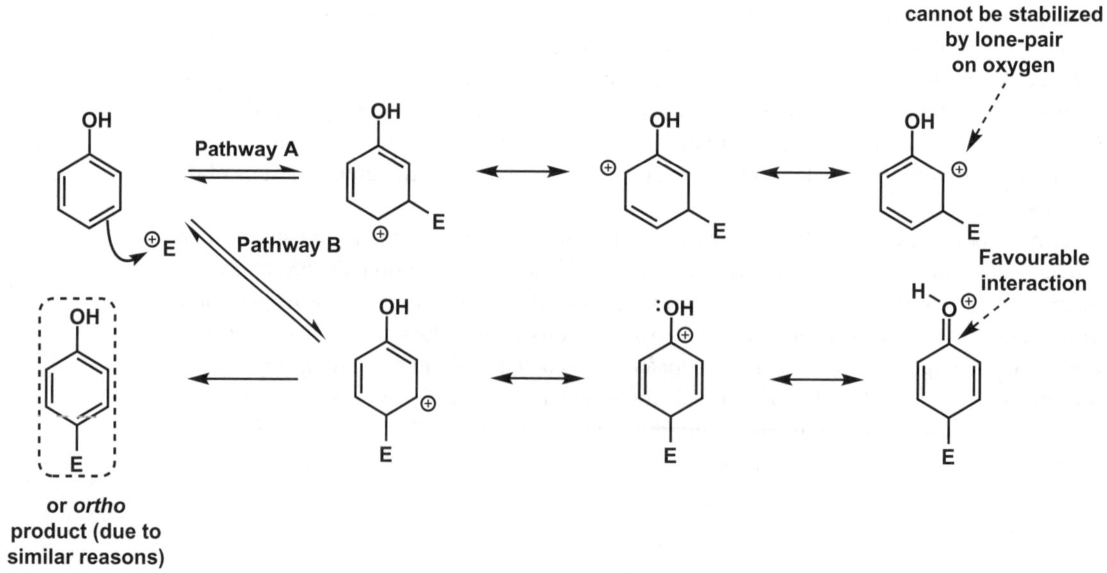

➜ For a refresher on resonance, see Chapter 1, section 1.8.

Worked example 6.2B

Suggest a mechanism for the following reaction. Account for the two products.

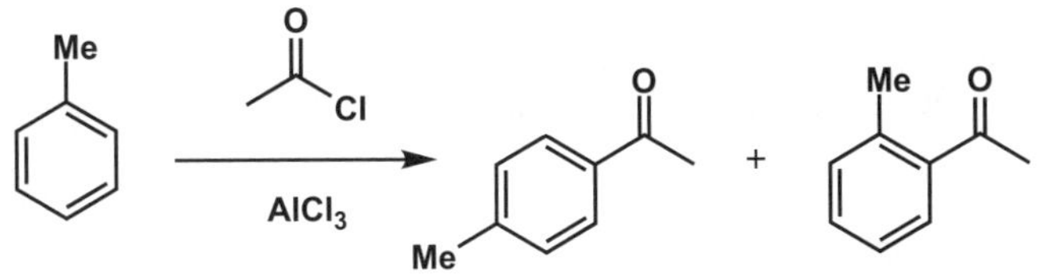

[Reaction scheme: toluene + acetyl chloride, AlCl₃ → 4′-methylacetophenone (para) + 2′-methylacetophenone (ortho)]

Solution

There are three clues in this question that will help you to determine the mechanism of this reaction.

1. Both AlCl₃ and an acyl chloride are present. Both of these things are required for a Friedel–Crafts reaction, suggesting an S_EAr mechanism.
2. The methyl group is an electron-donating group so the benzene ring is electron-rich, giving further evidence towards an S_EAr mechanism.
3. There are two products; one where the acyl group is added *ortho* and one where the acyl group is added *para* to the methyl group. This final piece of evidence suggests that an S_EAr reaction pathway has been followed.

Acetyl chloride is not a strong enough electrophile to undergo S_EAr without some help. In this case, the AlCl₃ activates the starting material by coordinating to the chlorine which allows generation of an acylium ion. This acylium species is now electrophilic enough to react with the aromatic ring.

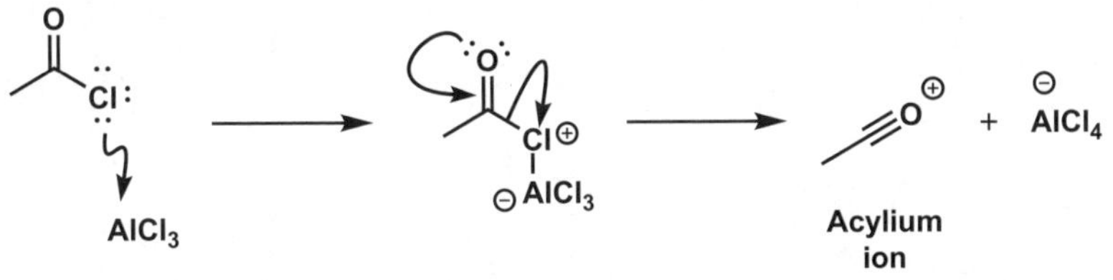

[Mechanism: acetyl chloride + AlCl₃ → acetyl–Cl–AlCl₃ complex → acylium ion + AlCl₄⁻]

The reason there are two products is because the methyl group is electron inducing and, if we ignore any steric effects, the ring can be acylated either *ortho* or *para* to the methyl group. Once the Wheland cation has been generated by attack of the aromatic ring, it can be placed on the carbon adjacent to the methyl group; a stabilizing interaction. In this case, we can assume that both the *ortho* and *para* products are formed with equal probability as no ratio has been given.

→ Sometimes it can be helpful to think of the *ortho* and *para* positions as having a slight negative charge due to hyperconjugation and inductive effects. For further information about these effects, see Clayden et al. (2012).

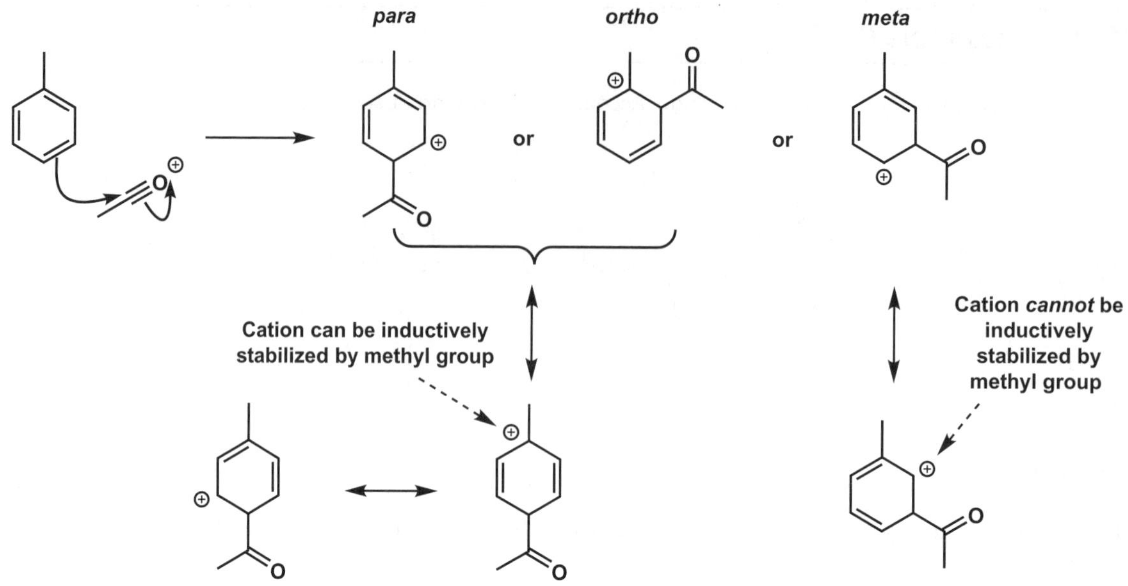

6 AROMATIC CHEMISTRY

para — Cation can be inductively stabilized by methyl group

ortho — Cation can be inductively stabilized by methyl group

meta — Cation *cannot* be inductively stabilized by methyl group

❓ Question 6.4

Suggest why there are differences in how many times the bromine is substituted onto the rings in the examples below.

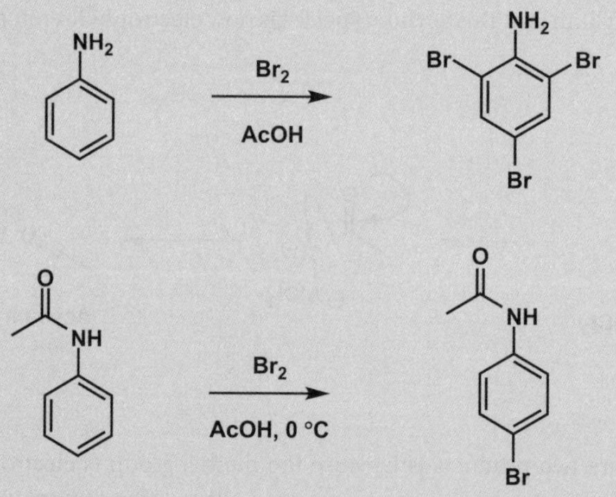

aniline + Br$_2$ / AcOH → 2,4,6-tribromoaniline

acetanilide + Br$_2$ / AcOH, 0 °C → 4-bromoacetanilide

▶ **Hint** Consider mesomeric and inductive effects.

Question 6.5

Suggest which of these reactions will proceed more quickly and why. Suggest a mechanism for each.

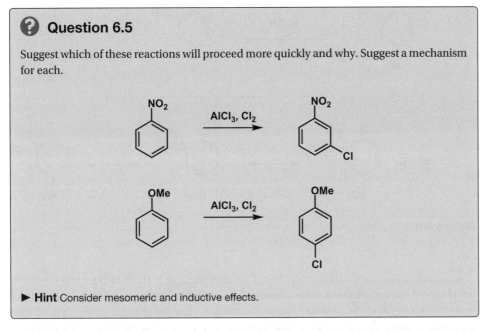

▶ **Hint** Consider mesomeric and inductive effects.

Question 6.6

Suggest the product or products of the reaction below. Which position is most likely to be substituted?

6.3 Nucleophilic aromatic substitution

What is nucleophilic aromatic substitution?

Nucleophilic aromatic substitution (S_NAr) is where a group on an aromatic ring is replaced by a nucleophile. It is the opposite of an electrophilic aromatic substitution, in that the ring acts as an electrophile rather than a nucleophile. This reaction requires a nucleophile to attack the aromatic ring, to generate an anion (Figure 6.3). The aromatic ring therefore must be electron-poor to both facilitate attack by the nucleophile, and also to stabilize the negative charge that is generated in the intermediate. In this reaction, the functional group or atom that is replaced must be a good leaving group (LG); H is rarely substituted. Common leaving groups are F, Cl, Br, and I. It is important that you do not confuse this type of reaction with $S_N 2$ substitution. $S_N 2$ substitution **cannot** occur at an sp^2 carbon; S_NAr follows an addition–elimination mechanistic pathway.

➔ For a refresher on nucleophiles, see Chapter 3, section 3.1.

➔ For a refresher on what makes a good leaving group, see Chapter 3, section 3.4.

Figure 6.3 Nucleophilic aromatic substitution mechanism.

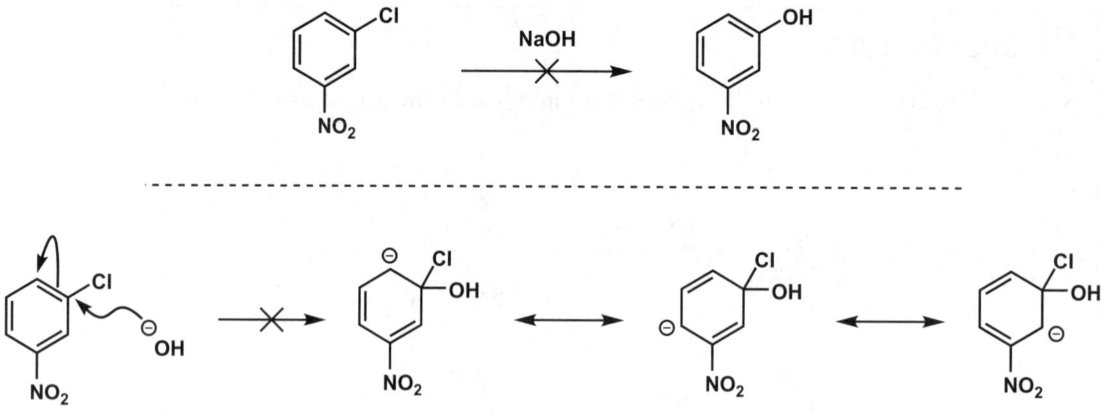

6 AROMATIC CHEMISTRY

Figure 6.4 Unsuccessful $S_N Ar$.

As stated earlier, this reaction works best if the aromatic ring is electron-poor and the resulting LG anion can be stabilized. If we consider the reaction in Figure 6.4, on initial inspection it looks feasible because the aromatic ring is electron-poor and the chloride atom is a good leaving group. However, this reaction is unlikely to take place readily because the anion generated cannot be stabilized onto the nitro group, i.e. the anion generated cannot be placed directly on the adjacent carbon atom.

Worked example 6.3 A

Draw the mechanism for the following reaction.

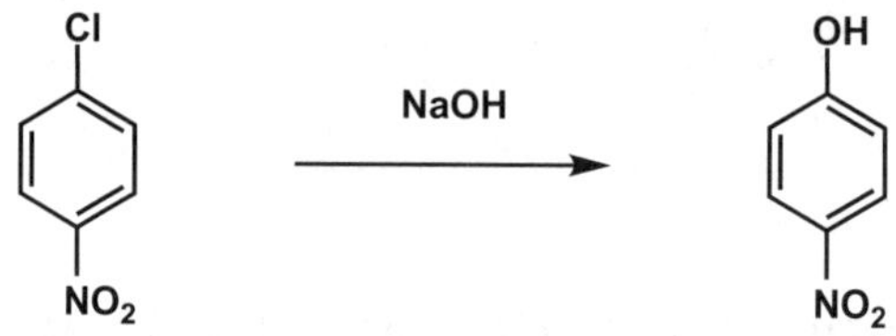

Solution

In this example, there is again a chlorine atom and a nitro group on the ring. The first thing to do is to look and see which has been replaced; in this example it is the chlorine, which is what we would expect as chlorine is a better leaving group than NO_2. The next thing to do is inspect the relationship of the substituents. In this example the nitro group is *para* to the chlorine which will be substituted. This is of benefit as the negative charge generated in the intermediate can be stabilized through resonance onto the nitro group and the reaction therefore proceeds.

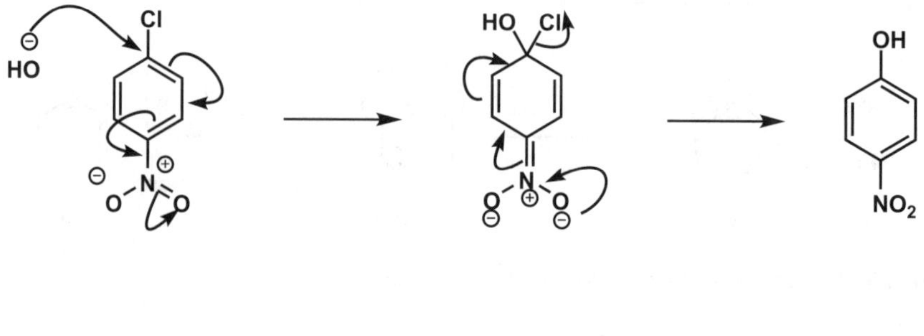

6.3 NUCLEOPHILIC AROMATIC SUBSTITUTION

Worked example 6.3B

Is the following reaction likely to occur?

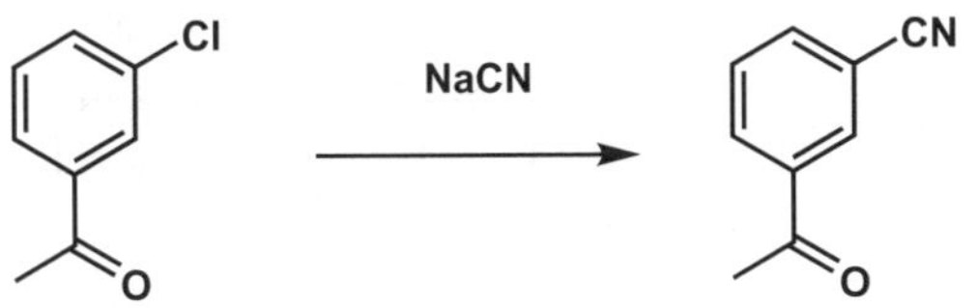

Solution

The first thing to check is the substituents on the ring; in this case there is a chlorine and a ketone. The chloride is a good leaving group, so can be substituted, and the ketone is electronically withdrawing hence the ring is electron-poor. Thus far, the conditions required for a S_NAr reaction have been fulfilled. The next step is to check the relationship between the leaving group and the EWG. If we attack the ring with the nucleophile, an anion is generated. However, this anion cannot be resonance stabilized by the ketone, so this reaction is unlikely to happen.

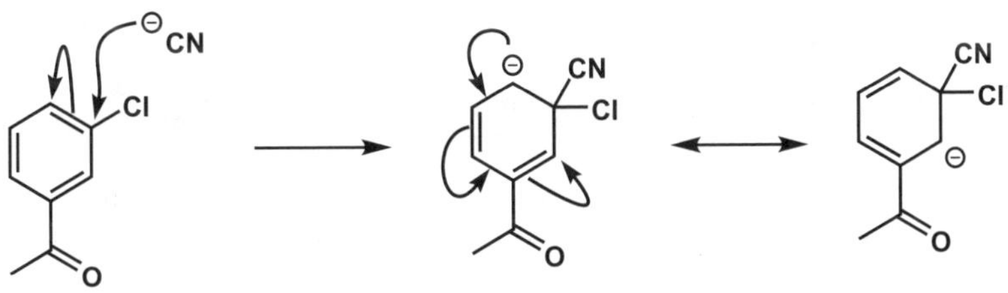

? Question 6.7

Suggest a mechanism for the reaction below. Why are two equivalents of the amine required?

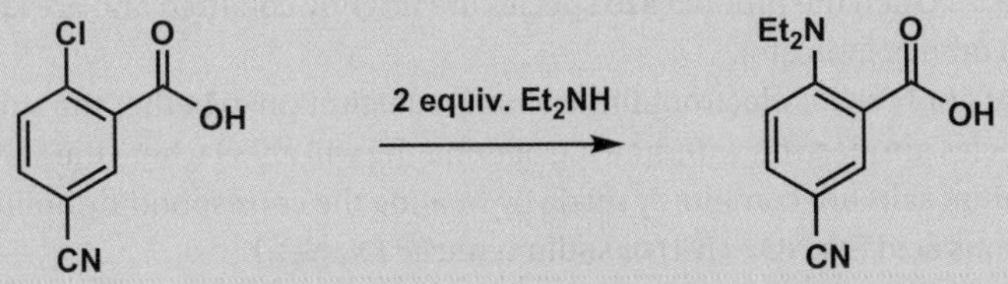

▶ **Hint** Think pK_a!

6 AROMATIC CHEMISTRY

> **Question 6.8**
>
> Suggest why only one chlorine is substituted in the reaction below.
>
>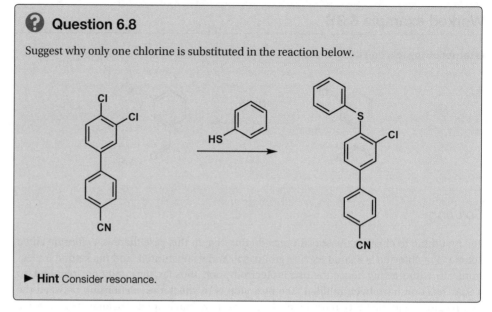
>
> ▶ **Hint** Consider resonance.

> **Question 6.9**
>
> Suggest a mechanism for the reaction below.

6.4 Azo coupling

What is azo coupling?

Azo coupling is when an aromatic diazonium-containing compound is joined to another aromatic compound. The mechanism involves attack of the diazonium group ($R-N_2^+$) by another aromatic ring therefore the nucleophile must be electron-rich to enable attack of the diazonium species (Figure 6.5). Often the product azo species are brightly coloured and are known as azo-dyes, e.g. methyl orange indicator.

The same rules apply as for electrophilic aromatic substitution; whether the substituents on the donating species are electron withdrawing or donating will affect where the substitution occurs. The diazonium salts are commonly made by treating the corresponding aniline (aromatic amine) with nitrous acid ($NaNO_2/HCl$) or sodium nitrite ($NaNO_2$).

Figure 6.5 Representative azo-coupling of a diazonium salt and an aromatic ring.

Worked example 6.4A

Suggest a mechanism for the following reaction.

[Reaction scheme: 4-methoxyphenol + phenyldiazonium cation → 2-(phenylazo)-4-methoxyphenol]

Solution

In this question the first thing to do is to inspect the functional groups on the aromatic ring and consider any electronic effects they may have. In this case there is a hydroxyl group and a methoxy group, both of which are *ortho/para* directing. Both of these groups render the ring electron-rich so it is able to act as a nucleophile and attack the azo-species.

In this example the lone-pair on the oxygen is used to push electron density into the ring, generating an oxonium intermediate. This intermediate then loses a proton to recover aromaticity, a highly favourable process.

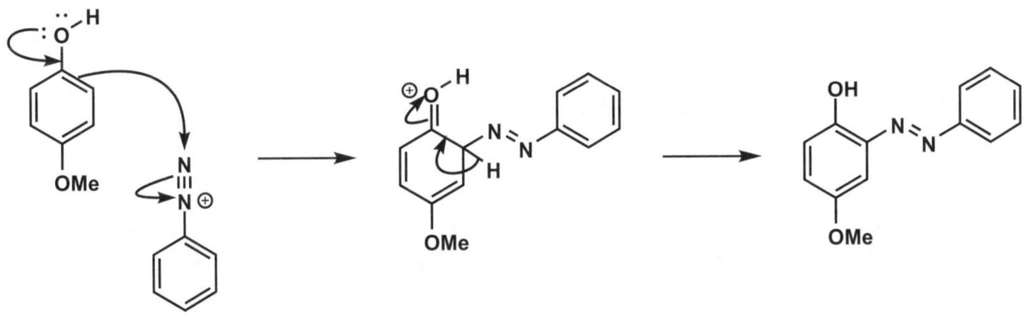

→ In this example the inductivity of OH dominates that of OMe. The reasons for this are beyond the scope of this textbook.

[Mechanism scheme showing arrow pushing from OH lone pair into ring, attack on diazonium, oxonium intermediate, then deprotonation to give product]

Worked example 6.4B

Suggest a mechanism for the following reaction.

[Reaction scheme: 2-hydroxybenzaldehyde (salicylaldehyde) + 4-nitrophenyldiazonium cation → azo coupling product]

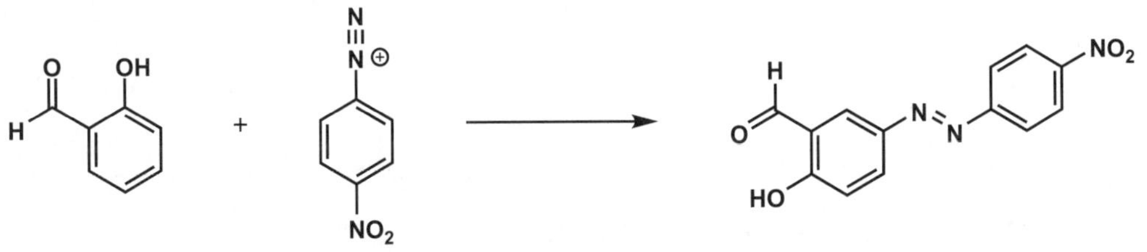

Solution

In this example there are again two directing groups; the hydroxyl group is electron donating, therefore activates the *ortho* and *para* positions towards nucleophilic attack. The aldehyde is electron-withdrawing so deactivates the *ortho* and *para* positions, but the *meta* position is unaffected. However, the hydroxyl group is an extremely strong donor therefore the deactivation effect from the aldehyde is over-ridden. As in the previous example, the oxygen lone-pair is used to push electron density onto the ring and then to attack the electrophile, generating an oxonium intermediate. This intermediate then loses a proton to regain aromaticity. Attack *para* rather than *ortho* is most likely due to steric reasons: the OH group is interfering with approach of the diazonium species.

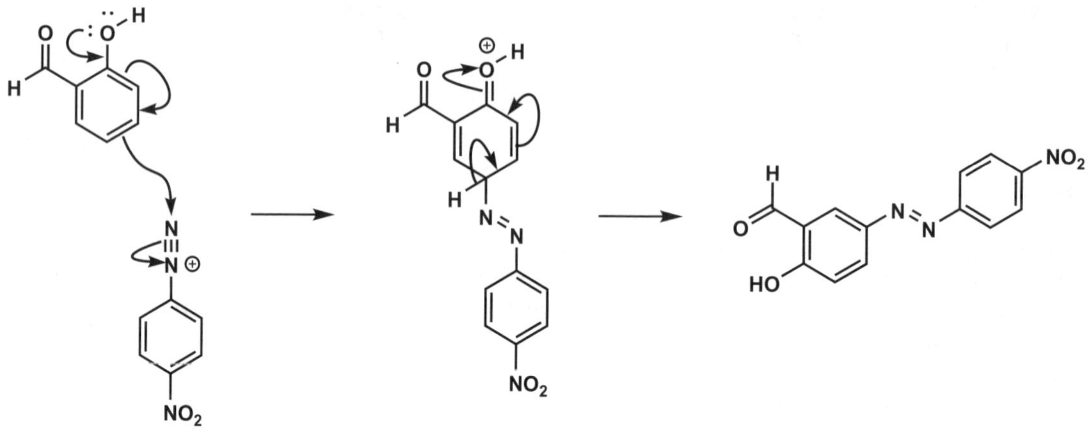

> **Question 6.10**
>
> Suggest a mechanism for the following reaction.
>
> ▶ **Hint** Look at the product and then work backwards. 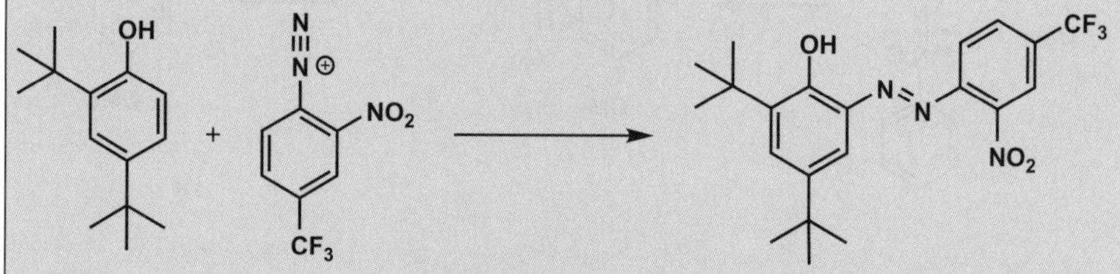

> **Question 6.11**
>
> Suggest a mechanism for the following reaction.

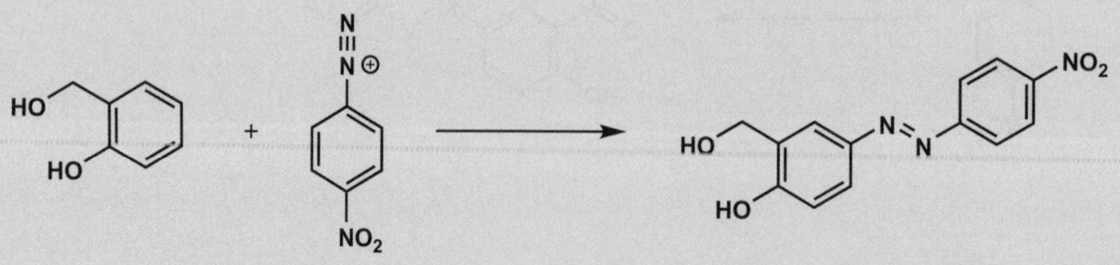

Question 6.12

Suggest a mechanism for the following reaction. Why does substitution occur in the position indicated?

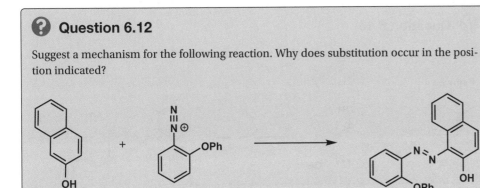

▶ **Hint** Consider resonance.

6.5 Synoptic questions

Question 6.13

Suggest the possible product(s) from the following reaction and provide a mechanism. Suggest which product is least likely and why.

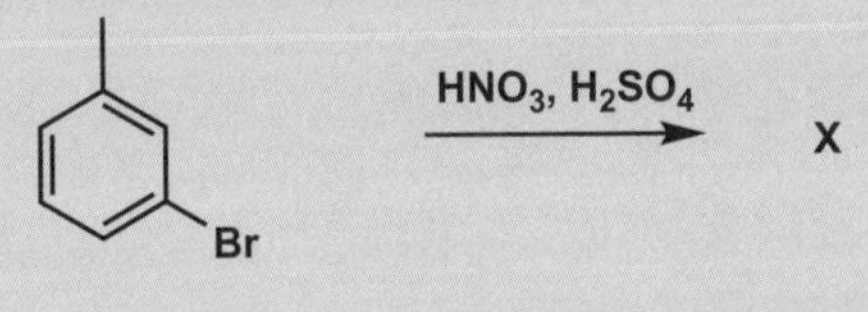

Question 6.14

Is the reaction below likely to occur? Give reasons for your answer.

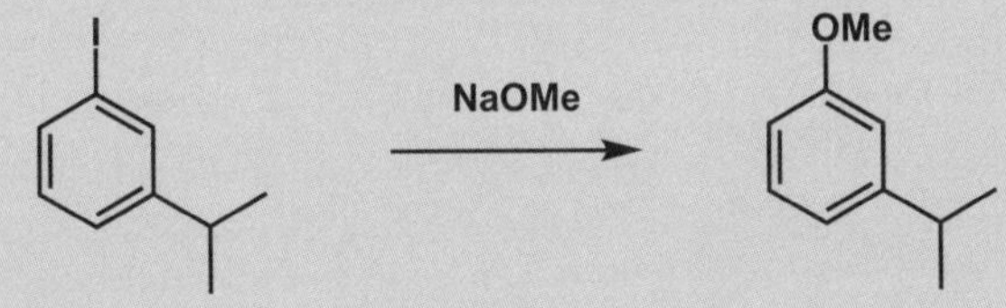

Question 6.15

How would you synthesize the following molecule, starting from benzene?

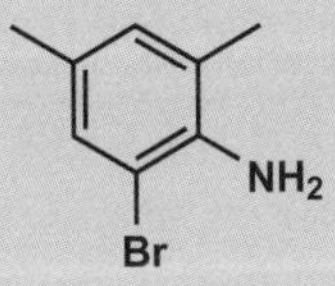

6 AROMATIC CHEMISTRY

> ## ❓ Question 6.16
>
> Suggest which synthetic approach would undergo faster reaction.
>
> Pathway A
>
> Pathway B
>
>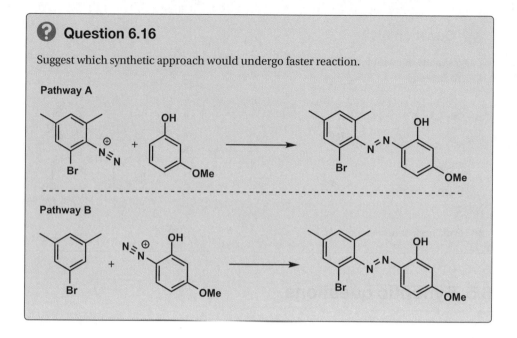

References

Clayden, J., Greeves, N., and Warren, S. (2012) *Organic Chemistry*, 2nd edn (Oxford University Press, Oxford).

7
Carbonyl chemistry

7.1 Structure and bonding

What is the carbonyl group?

In its simplest form, the carbonyl group is a carbon to oxygen double bond (Figure 7.1). Both the carbon and oxygen are sp² hybridized because they are joined by a double bond, and are therefore planar.

➔ For a refresher on hybridization, see Chapter 1, section 1.4.

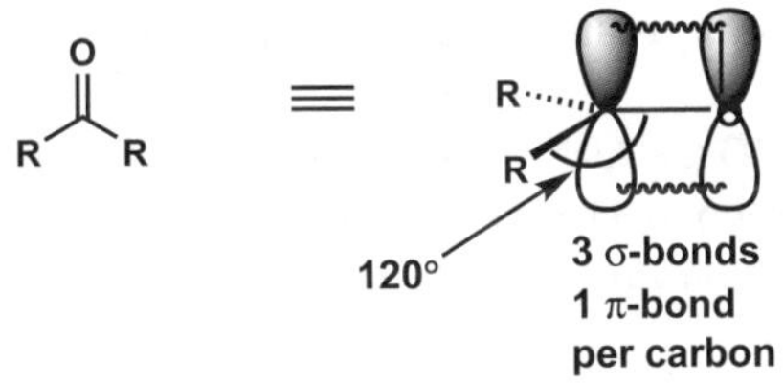

Figure 7.1 The carbonyl group.

A neutral carbon atom must have four bonds, therefore the nature of the other substituents at the central carbon atom will dictate what type of carbonyl group it is and also its reactivity profile.

There are numerous functional groups that contain a carbonyl moiety. Some of the groups that you are most likely to encounter are shown in Figure 7.2.

If we consider the electronic structure of the carbon–oxygen double bond, we know that oxygen is an electronegative element therefore the double bond is polarized towards the oxygen; there is a larger orbital coefficient (i.e. more electron density) on the oxygen in the π-bonding

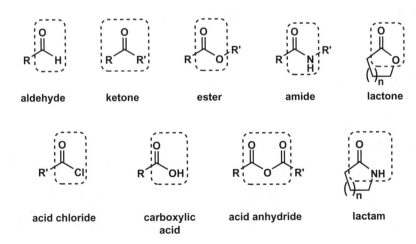

Figure 7.2 Carbonyl-containing functional groups.

7 CARBONYL CHEMISTRY

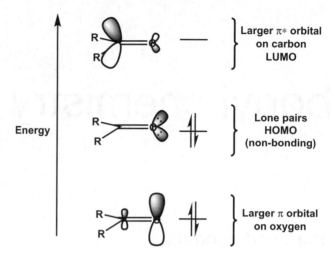

Figure 7.3 The relevant molecular orbitals of the carbonyl group.

→ For a refresher on bonding and anti-bonding, see Chapter 1, section 1.3.

orbital (Figure 7.3). When considering the anti-bonding orbital (π*) the opposite is true; the larger anti-bonding lobe is on the carbon. The lone pairs on oxygen are also important and are known as non-bonding. They are useful because they are the HOMO, i.e. the most available pair of electrons.

Consideration of the HOMO and LUMO orbitals is extremely important when discussing the reactivity of the carbonyl group; nucleophiles attack the carbonyl group at the π* orbital (the LUMO), i.e. at the central carbon atom, and electrophiles are attacked by the oxygen lone pairs (HOMO) (Figure 7.4).

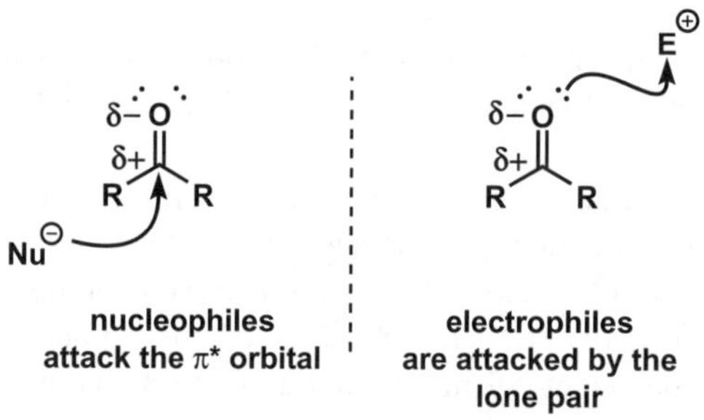

Figure 7.4 Reactivity modes of the carbonyl group.

Worked example 7.1A

Name the carbonyl groups in the following compounds:

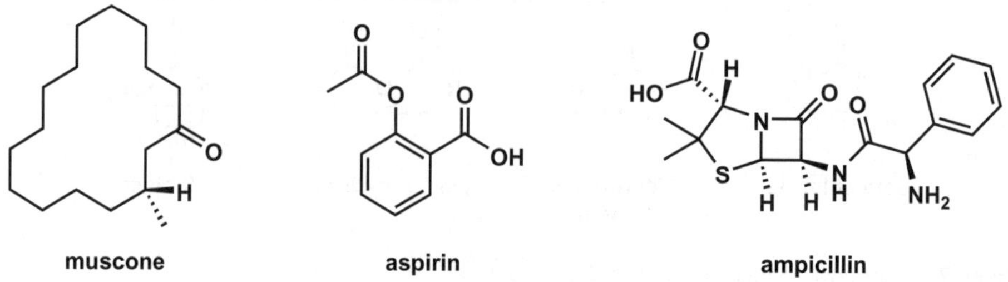

muscone   aspirin   ampicillin

Solution

In order to answer this question, you need to look at the carbon of the C=O and see what is either side of it. In muscone there is only a ketone; the central carbon atom has a carbon on either side. In aspirin there are two carbonyl groups, both of which are flanked by one oxygen and one carbon. In this case we now need to look and see what is bonded to the oxygen. If the oxygen is bonded to a hydrogen, making an OH group, the functional group is a carboxylic acid. If the oxygen is bonded to another carbon atom, the functional group is an ester. Finally, in ampicillin there are three C=O containing moieties; in one the central carbon is bonded to an oxygen and a carbon, and in the other two the central carbon is bonded to a carbon and a nitrogen. The first oxygen-connected moiety is a carboxylic acid, for reasons discussed previously. The nitrogen containing moieties are both amides; however, if an amide is contained within a ring, it is referred to as a lactam. In this case it is a β-lactam because the amide-containing ring has four atoms.

→ A γ-lactam ring contains five atoms and a δ-lactam ring contains six atoms in total.

Worked example 7.1B

By comparing the structures of an aldehyde, a ketone, an ester, and an amide, which do you think has the largest partial positive charge on carbon and why? How will this affect the reactivity?

Solution

The aldehyde and ketone are both flanked by atoms which do not have any lone pairs of electrons (C and H) whereas in the ester and amide one of the flanking atoms has a lone pair of electrons. The lone pair of electrons on both the oxygen and the nitrogen can be donated into the carbonyl group, therefore increasing the electron density on the central carbon. This means that the central carbon is less reactive towards nucleophiles because it is less δ+. We can now make a judgement that the aldehyde and ketone are more reactive than the ester and the amide.

Looking more closely at the aldehyde and the ketone, we can see that the aldehyde has an adjacent carbon and a hydrogen, and the ketone has two carbons. This makes a large difference to their comparable reactivity. The alkyl chains on the ketone are important, first because they are able to inductively push electron density towards the central carbon atom (known as hyperconjugation), and second because they are sterically large and hinder any incoming nucleophile.

→ For more information about the hyperconjugation effect of alkyl chains and ketones, see Clayden et al. (2012).

> For a refresher on the inductive effect, see Chapter 1, section 1.6.

The aldehyde only has one alkyl chain that is able to offer this induction effect, therefore the stabilization of the δ+ charge is lower, so the aldehyde is more reactive towards nucleophiles because the central carbon is slightly more δ+ than that of the ketone. Additionally, the steric bulk of one H and one alkyl chain is less than that of two alkyl chains.

> For a more detailed explanation, see Clayden et al. (2012).

We now need to compare the reactivity of the ester and the amide, both of which are less reactive than the ketone and aldehyde due to the lone pairs on the adjacent heteroatoms (O and N). The ester contains an adjacent oxygen and the amide an adjacent nitrogen; both electronegative elements. Oxygen is more electronegative than nitrogen and therefore less able to donate its lone pair of electrons into the carbonyl group so the amide has a smaller δ+ charge on the central carbon compared to the ester. If we compare the energy of the lone pairs on O and N we can see why this is: the oxygen lone pair is lower in energy due to the electronegativity of oxygen, therefore orbital overlap with the high-energy π* orbital of the C=O bond is less likely. This means that when comparing esters and amides, esters are more reactive.

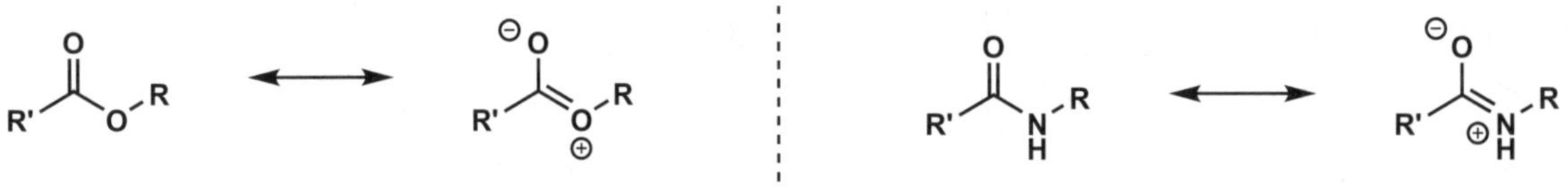

The final reactivity series is therefore (most reactive first) aldehyde > ketone > ester > amide.

> ### ❓ Question 7.1
>
> Do you think that cyclopropanone (shown below) is very stable to nucleophiles? Give some reasons to support your argument. 
>
>
>
> ▶ **Hint** Consider the bond angles.

> ### ❓ Question 7.2
>
> Which carbonyl group in the series below do you think will have the largest partial positive charge on carbon and why? 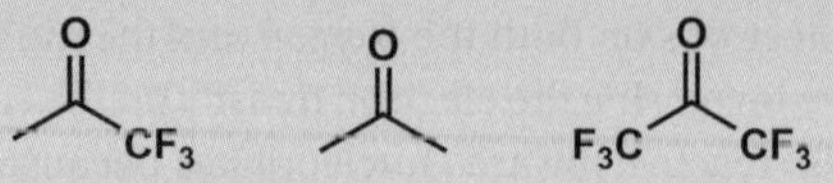
>
>
>
> ▶ **Hint** Consider inductive effects.

Question 7.3

We usually depict the ester group as shown below with the C—C and O—C bonds aligned with each other. Suggest four reasons for why this is the correct way of drawing the ester group.

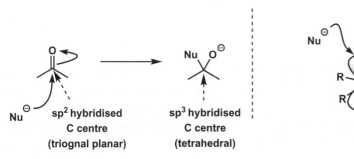

aligned with each other

7.2 Reactions with nucleophiles

Where does the nucleophile react?

As discussed earlier in the chapter (Figure 7.3), a carbonyl group has a larger orbital coefficient on oxygen in the π-bonding orbital. When considering the anti-bonding orbital, the π*, the opposite is true; the larger anti-bonding lobe is on the carbon. The lone pair of electrons on the oxygen are the HOMO, and the π* anti-bonding orbital is the LUMO. These are the most important orbitals when discussing the reactivity of the carbonyl group with both nucleophiles and electrophiles.

Nucleophiles contain either a negative charge or a readily available pair of electrons that can be donated. A nucleophile always has to react with a lowest energy empty orbital, the LUMO (Figure 7.5). In the case of a carbonyl group, the lowest energy empty orbital is the π*, which has its largest orbital coefficient on the carbon. Once the carbonyl group has been attacked, the hybridization of the central carbon changes from sp² to sp³ and therefore the shape changes from trigonal planar to tetrahedral. The angle of attack by the nucleophile is 107° in relation to the C=O bond and corresponds to the location of the π* orbital. This angle is known as the Bürgi–Dunitz angle or Bürgi–Dunitz trajectory.

If there is a leaving group on the carbonyl, e.g. Cl, the anionic oxygen generated from the nucleophilic addition can eliminate the leaving group to reform the C=O bond, therefore regenerating an sp² hybridized carbon (Figure 7.6). This may be the final product, or, depending upon the reaction conditions, it is possible for this newly formed carbonyl group to react further with another nucleophile.

→ Similar to other chemistry discussed in this book, the ability of the leaving group to leave is related to the pK_a of the parent acid. For more information, see Chapter 3, section 3.4.

Figure 7.5 Nucleophile reactivity general scheme.

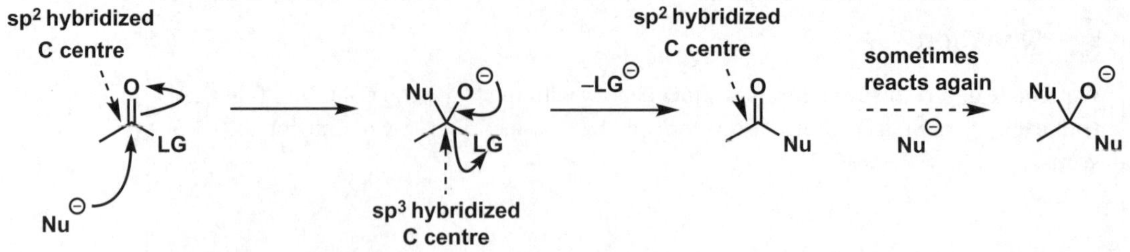

7 CARBONYL CHEMISTRY

Figure 7.6 Nucleophile reaction with a leaving group.

For a more information, see Clayden et al. (2012).

Alternatively, if the incoming nucleophile has a free electron pair after it has attacked the C=O, this electron pair can be used to eliminate water (Figure 7.7). Examples of elements which commonly do this are oxygen, to generate an oxonium species that can undergo further reaction, or nitrogen to generate imines (or enamines) as shown. Usually these reactions require an acid catalyst.

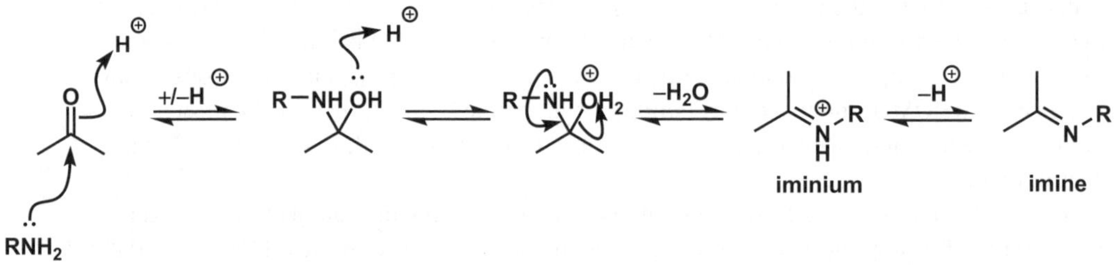

Figure 7.7 Imine formation resulting from elimination of water.

Worked example 7.2A

Suggest the mechanism of the following reaction.

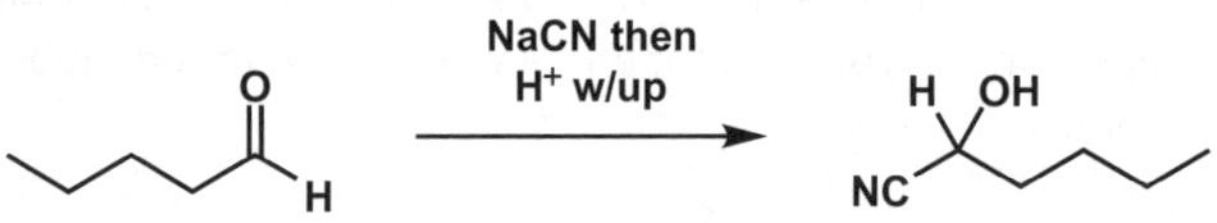

Solution

The key to answering this question is first to identify the nucleophile and use it to attack the electrophile (LUMO, C=O π^*). NaCN, sodium cyanide, is an ionic solid which exists as a sodium cation and a cyanide anion. The cyanide anion can act as a nucleophile because it has a discrete negative charge and will attack the C=O π^*, generating a tetrahedral oxy-anion that can be quenched upon work-up to generate a cyanohydrin.

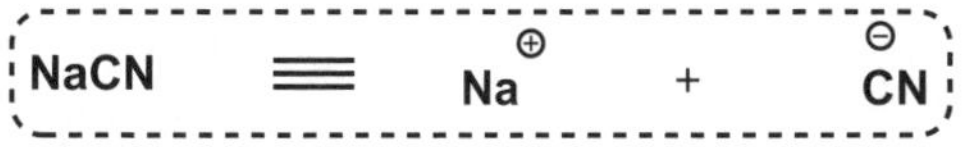

7.2 REACTIONS WITH NUCLEOPHILES

NaCN ≡ Na⁺ + CN⁻

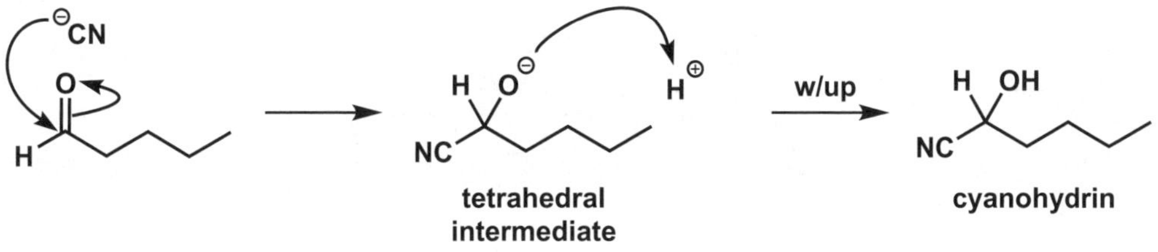

[Mechanism showing CN⁻ attacking aldehyde carbonyl to form tetrahedral intermediate, then protonation on workup (w/up) to give cyanohydrin.]

tetrahedral intermediate → cyanohydrin

Worked example 7.2B

Suggest a mechanism for the following reaction, assuming there are at least two equivalents of ethyl magnesium bromide present. Suggest why no intermediate X is isolated if only one equivalent of nucleophile was used.

[Ethyl butanoate + EtMgBr, then H⁺ → X → tertiary alcohol (3-ethyl-3-hexanol type)]

Solution

For the magnesium species the carbon can be thought of as carrying a negative charge.

The negative charge is a nucleophile and will therefore attack the low-lying π* orbital of the C=O bond, generating an oxy-anion. Due to the leaving group on the carbonyl carbon (OEt), the negative charge generated will eliminate ethoxide, to make a ketone which is more reactive than the starting ester (see section 7.1). This ketone then undergoes nucleophilic attack to generate a final alkoxide species that cannot eliminate anything. This is because the leaving group has to be a carbanion, having a higher pK_{aH} than the alkoxide species (48 in the case of the carbanion versus 17 for the ethoxide), and is therefore a very unfavoured process.

If only one equivalent of the magnesium species were added we would most likely end up with a 1:1 mixture of starting ester and tertiary alcohol. This is because the ketone intermediate is more reactive than the ester starting material so the magnesium reagent will react with the ketone more readily than the ester.

➔ Organomagnesium species like the one shown are called 'Grignard reagents' and are very useful. You will learn more about them at a later stage in your degree course. For more information, see Clayden et al. (2012).

➔ Elimination is only possible if the pK_{aH} of the eliminated species is low. Ethoxide can be eliminated because ethanol, the protonated species, has a pK_a of approximately 17. However, to eliminate an alkyl chain would give a species with a pK_a of around 48. For a recap relating pK_a to leaving group ability, see Chapter 3, section 3.4.

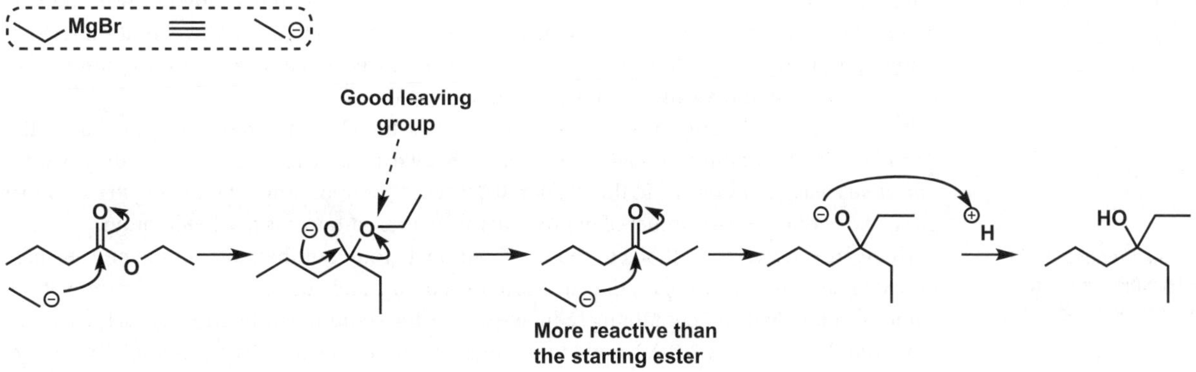

EtMgBr ≡ Et⁻

[Mechanism: ester attacked by Et⁻, tetrahedral intermediate with OEt as good leaving group, forms ketone (more reactive than the starting ester), second equivalent of Et⁻ attacks, protonation gives tertiary alcohol.]

Good leaving group

More reactive than the starting ester

Question 7.4

Suggest a mechanism and the product of the following reaction.

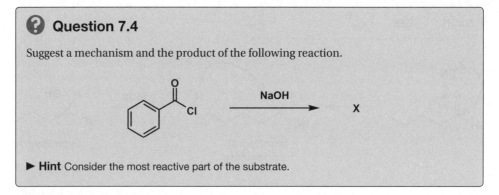

▶ Hint Consider the most reactive part of the substrate.

Question 7.5

Suggest a mechanism for the following reaction.

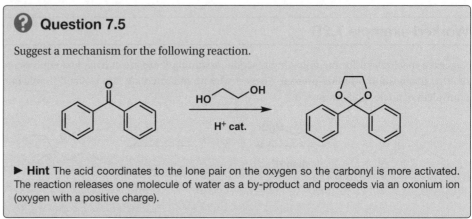

▶ Hint The acid coordinates to the lone pair on the oxygen so the carbonyl is more activated. The reaction releases one molecule of water as a by-product and proceeds via an oxonium ion (oxygen with a positive charge).

Question 7.6

Suggest a mechanism for the following reaction.

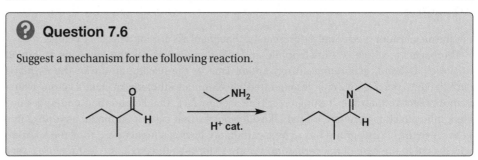

7.3 Reactions with reducing agents

Where does the reducing agent react?

→ For more information, see Clayden et al. (2012).

There are many different levels of oxidation state for carbon atoms ranging from the carbon dioxide oxidation level to the alkane oxidation level (Table 7.1). The oxidation level at which a particular carbon resides is related to the degree of saturation and the number of bonds it has to heteroatoms (i.e. an atom that is not C or H).

If X is O, the number of oxidation states with a carbonyl group present is quite large. The defined reactivity of the carbonyl group means that we are able to access most of these oxidation states reasonably easily. In this section we discuss common ways of reducing the carbonyl group, i.e. reducing the number of bonds that the central carbon has to a heteroatom.

→ Do not confuse these reagents with sodium hydride (NaH) which is not a reducing agent but a strong base.

There are a range of methods to reduce a carbonyl group but the most simple ones usually provide a source of hydride, H^-, that acts as a nucleophile and adds to the central carbon atom. Some common hydride-containing reducing agents are sodium borohydride ($NaBH_4$), sodium cyanoborohydride ($NaCNBH_3$), and lithium aluminium hydride ($LiAlH_4$). There are two others which you may also encounter, di*iso*butylaluminium hydride (DIBALH) and borane (BH_3), but

Table 7.1 The various oxidation levels of carbon. X denotes a heteroatom.

Oxidation Level	Alkane	Alcohol	Aldehyde	Carboxylic acid	Carbon dioxide
No of bonds to heteroatoms	0	1	2	3	4

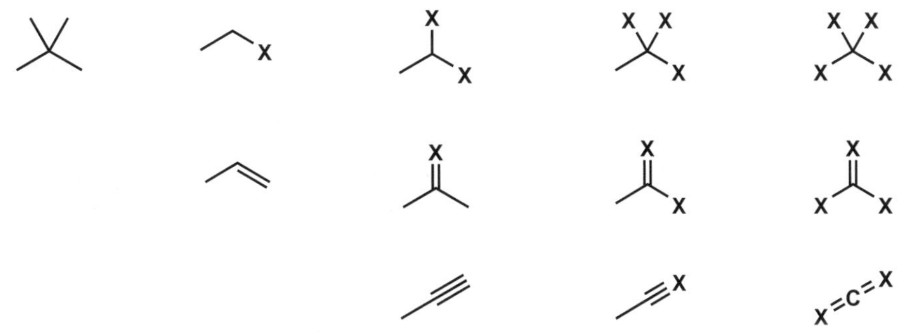

Table 7.2 General reactivity guide for different hydride-containing reducing agents.

		Increasing reactivity (from left to right)				
		aldehyde	ketone	ester	amide	carboxylic acid
Increasing reactivity (from top to bottom)	NaCNBH$_3$	maybe	maybe	✗	✗	✗
	NaBH$_4$	✓	✓	✗	✗	✗
	LiBH$_4$	✓	✓	✓	✗	✗
	DIBALH	✓	✓	✓	✓	✓
	LiAlH$_4$	✓	✓	✓	✓	✓

Smith, M. B. (2013) *March's Advanced Organic Chemistry: Reactions, Mechanisms, and Structure*, 7th edn (John Wiley & Sons Inc., Hoboken, NJ).

these reagents are slightly more advanced as they are require the hydride ion to be generated first through reaction of the substrate or solvent with the reducing agent.

The reducing agent that you use is important as they all react slightly differently. In the series above, sodium cyanoborohydride can be considered the mildest reducing agent and lithium aluminium hydride the strongest. The implication of this is that different reducing agents are best for different functional groups (Table 7.2). It is even possible to reduce one carbonyl-containing functional group in the presence of another, provided you choose an appropriate hydride source.

Worked example 7.3A

Suggest a mechanism for the following transformation.

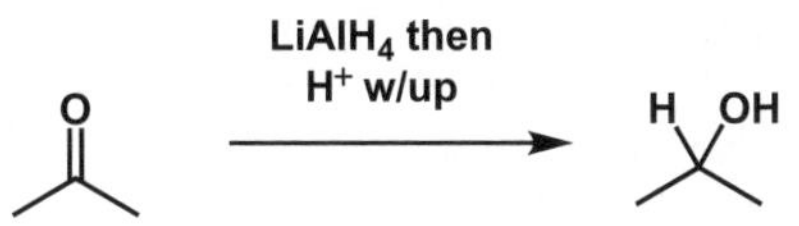

7 CARBONYL CHEMISTRY

Solution

→ Although this is a ketone, it is at the aldehyde oxidation level.

Lithium aluminium hydride (LiAlH$_4$) is a very reactive source of hydride. The hydride ion attacks the carbonyl to give the reduced product; the oxidation level of the central carbon has changed from the aldehyde to the alcohol oxidation level.

In this example, all that you need to appreciate is that a hydride is present and it will react in much the same way as a nucleophile because it has a negative charge (HOMO) which can react with the C=O π* orbital (LUMO).

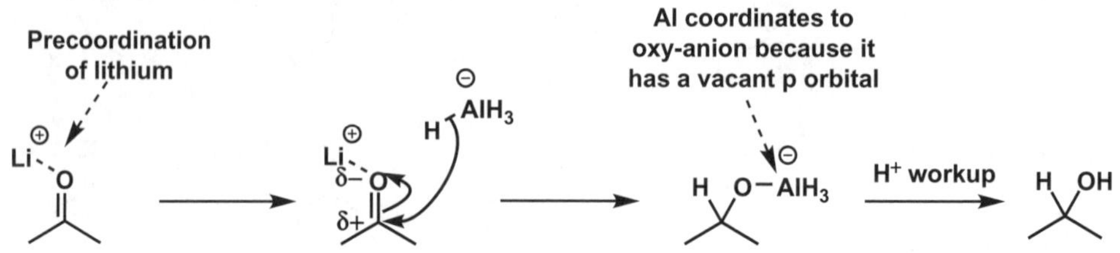

Worked example 7.3B

Suggest a mechanism for the following transformation.

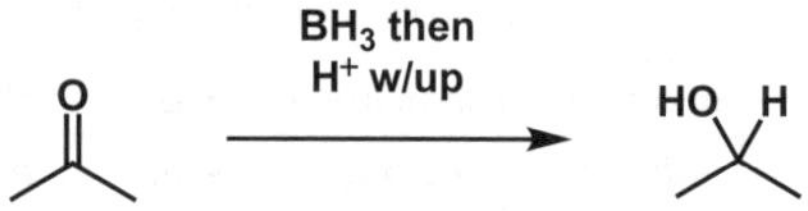

Solution

→ Warning: This is a very simplified version of this reaction mechanism!

When there is not hydride already present as H$^-$, e.g. in the case of DIBALH and borane (BH$_3$), the hydride has to be generated. This is achieved by coordination of the substrate to the reducing agent to form the active hydride species. In this example, the lone pair on the oxygen of the carbonyl group acts as a Lewis base, and can be donated into the empty p orbital on the boron generating a boronate species—the active reducing species. The hydride species required for reduction has now been generated and the hydride can be delivered to the carbonyl group intramolecularly, which is favourable. Additionally, boronate ester formation activates the carbonyl group (by acting as a Lewis acid) and the oxygen now has a positive charge so the π orbital is polarized and can undergo reduction much more easily. Acidic work-up provides the alcohol product.

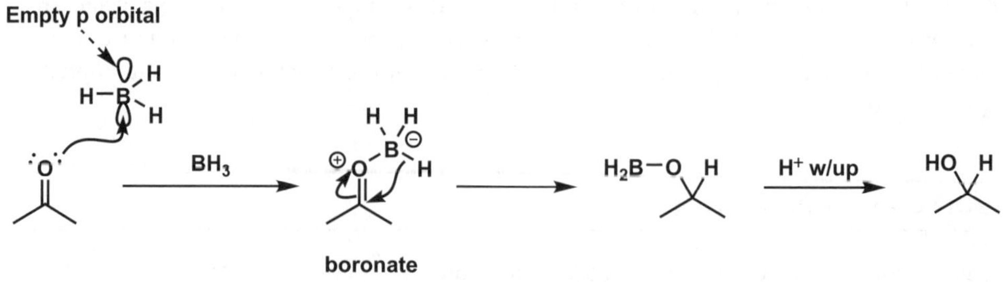

7.4 CARBOXYLIC ACIDS

> **Question 7.7**
>
> In the molecule shown below, suggest which carbonyl group would be reduced first by using NaBH$_4$ and why. 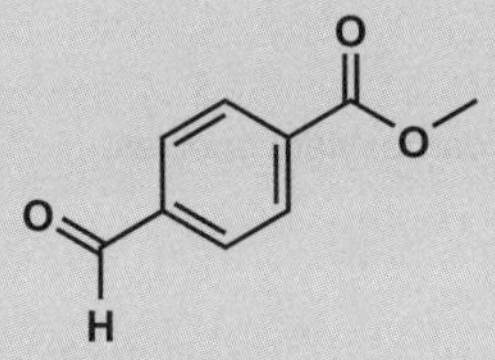
>
>
>
> ▶ **Hint** Consider electron-density at the reactive centre.

> **Question 7.8**
>
> What will be the product from one equivalent of DIBALH in this reaction? Provide the mechanism.
>
>

> **Question 7.9**
>
> Suggest a mechanism for the following reaction.
>
>
>
> ▶ **Hint** Consider the lone pairs on nitrogen. 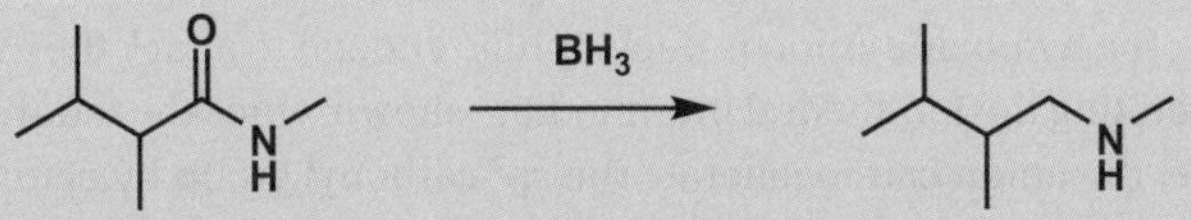

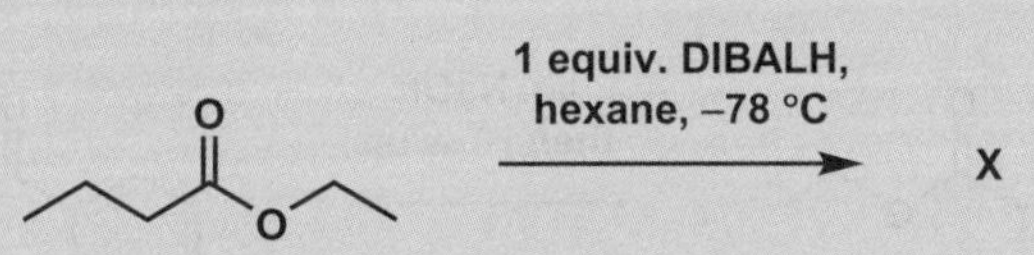

7.4 Carboxylic acids

What is a carboxylic acid?

Carboxylic acids are a class of carbonyl compounds that contain a carboxyl group. The terminal proton is reasonably acidic and has a pK_a of approximately 5, therefore it can be removed fairly easily.

When the proton has been removed and the acid is in the salt form, it is referred to as the 'carboxylate'. A carboxylate is not a very good nucleophile because the negative charge is spread over three atoms and is more diffuse (Figure 7.8).

⮕ For a refresher on pK_a, see Chapter 3, section 3.4.

How are they prepared?

Carboxylic acids can be prepared in a variety of ways. The most common are ester hydrolysis, amide hydrolysis, and by the oxidation of primary alcohols.

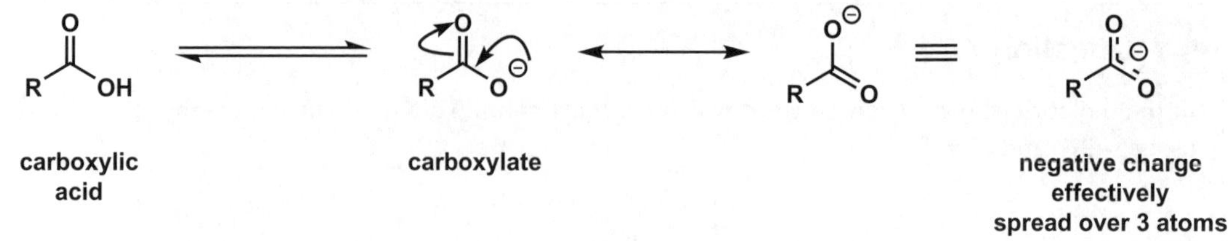

Figure 7.8 Resonance structures of a carboxylate species.

How do they react?

Carboxylic acids are reasonably versatile and can be used to prepare a variety of other functional groups. They can be converted into esters, acid chlorides, alcohols, and can also be alkylated at carbon to generate branched carboxylic acids.

→ For a refresher on reduction, see section 7.3.

Worked example 7.4A

Suggest a mechanism for the following reaction.

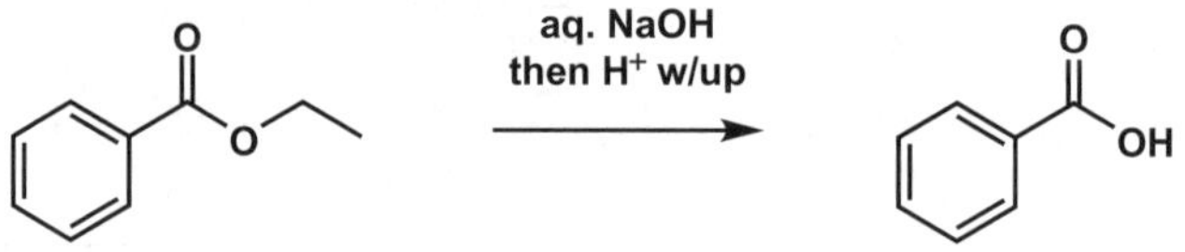

Solution

Sodium hydroxide is an ionic compound and in aqueous media exists as a sodium cation and a hydroxide anion. The hydroxide anion is nucleophilic enough to attack the ester at the δ+ carbonyl carbon, generating the tetrahedral intermediate shown. Once the tetrahedral intermediate has been made, the anion can regenerate the sp² carbonyl group by eliminating ethoxide. The reason it has been drawn as a carboxylate rather than a carboxylic acid is because the pK_a of hydroxide is 15 and that of a carboxylic acid 5, therefore the hydroxide ion will deprotonate the carboxylic acid as soon as it is formed because it generates a more stable anion. Only upon work-up with a strong acid, e.g. aq. HCl, will the carboxylate be protonated to generate the desired carboxylic acid product.

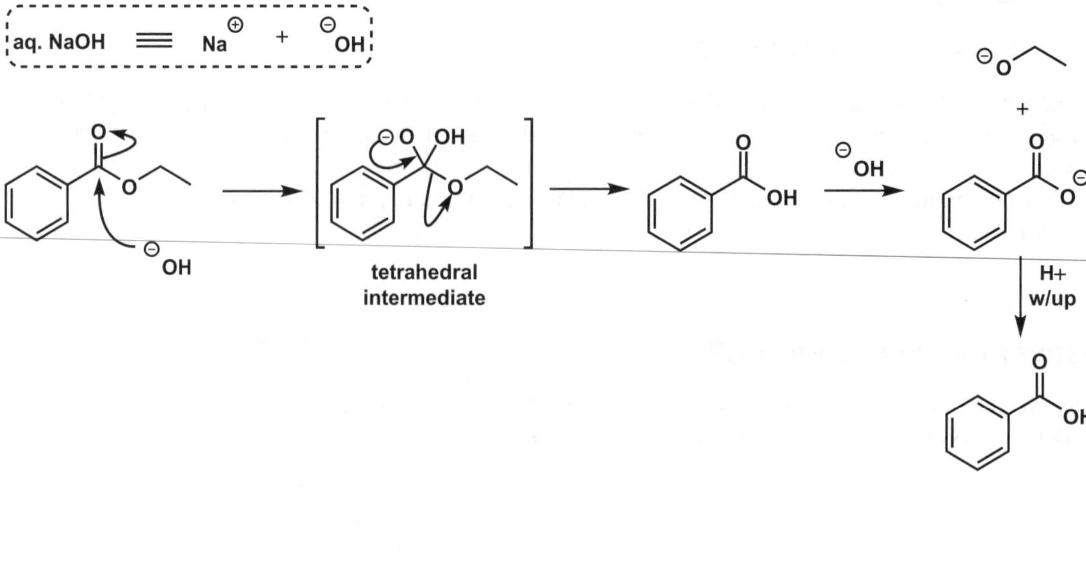

Worked example 7.4B

Suggest a mechanism for the following reaction.

PhCOOH + EtOH, cat. conc. H_2SO_4 → PhCOOEt

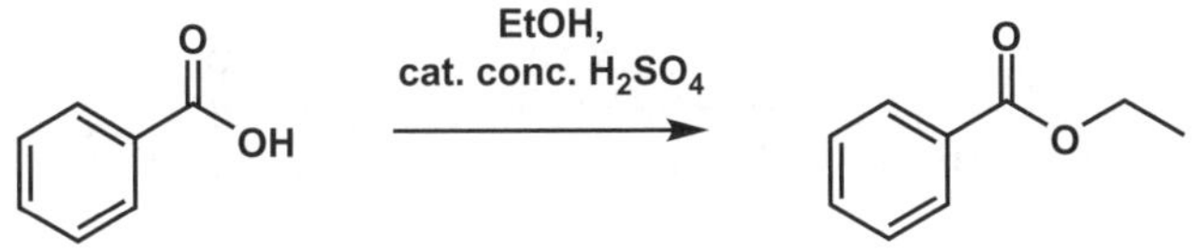

Solution

This reaction is known as a Fischer esterification and is the opposite of the other example discussed previously. The first step is protonation of the carbonyl group which activates it by causing an even bigger dipole so the δ+ charge on carbon is enhanced. This then allows the ethanol to attack to generate the intermediate **I**. Switching protons to generate a new oxonium species leads to intermediate **II**, which can eliminate water. Finally, deprotonation of intermediate **III** leads to the product. It is important to notice that all of the reaction steps are reversible therefore the whole system is in equilibrium.

→ This reaction is under acidic conditions therefore under no circumstances can hydroxide (HO⁻) be eliminated. It **must** be eliminated as water.

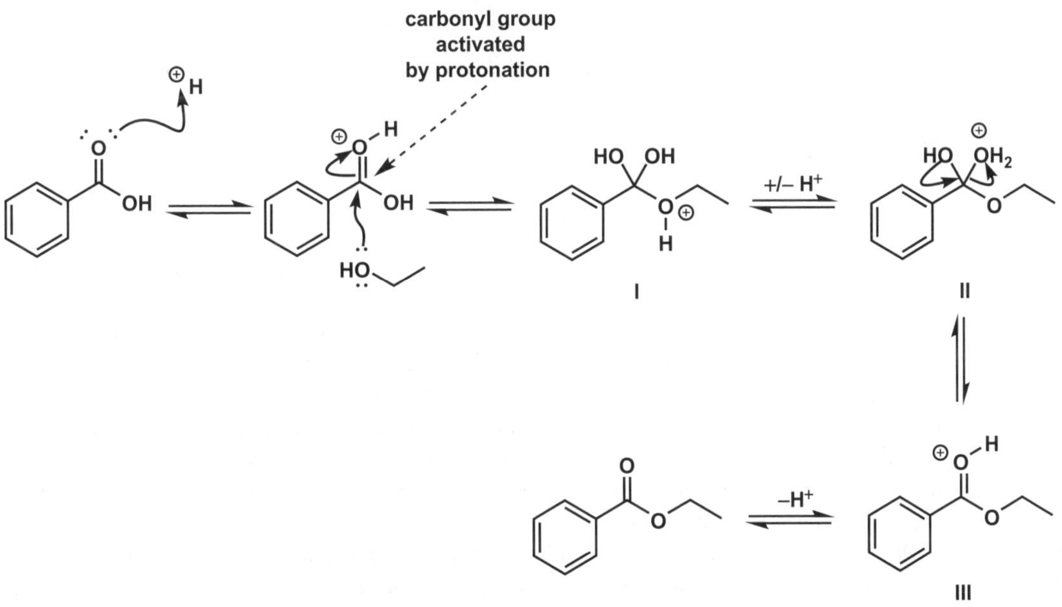

[Mechanism scheme showing: protonation of carbonyl group activated by protonation; ethanol attack to form intermediate I; proton switching (+/– H⁺) to intermediate II; loss of water to intermediate III; deprotonation (–H⁺) to give the ethyl benzoate product.]

? Question 7.10

Suggest a mechanism for the following reaction.

CH₃CH₂CH₂COOH + SOCl₂ → CH₃CH₂CH₂COCl + SO₂ + HCl

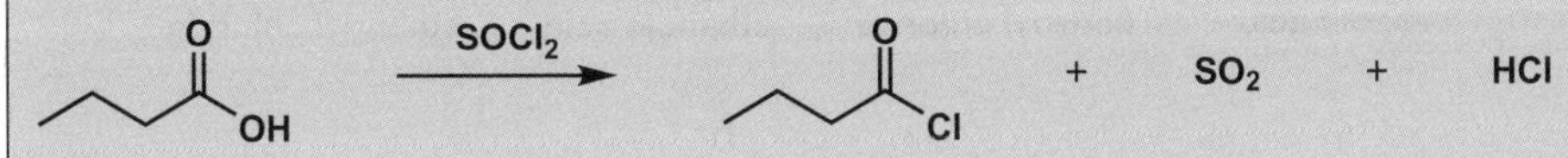

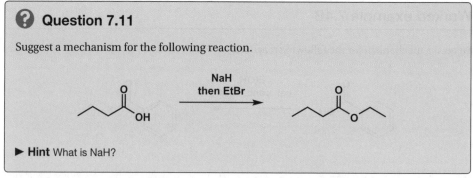

> **Question 7.11**
>
> Suggest a mechanism for the following reaction.
>
> ▶ **Hint** What is NaH?

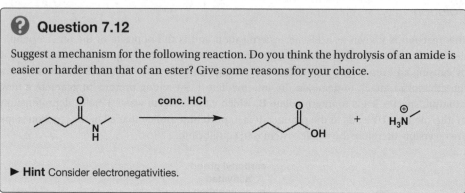

> **Question 7.12**
>
> Suggest a mechanism for the following reaction. Do you think the hydrolysis of an amide is easier or harder than that of an ester? Give some reasons for your choice.
>
> ▶ **Hint** Consider electronegativities.

7.5 Acyl chlorides

What is an acyl chloride?

Acyl chlorides are a class of carbonyl compounds where the central carbonyl carbon is bonded to a chlorine atom and another carbon. Acyl chlorides are extremely reactive and are used to generate a variety of other functional groups. Acyl chlorides are very prone to hydrolysis to the corresponding carboxylic acid. This is because the chlorine atom is electronegative, therefore pulls electron density towards itself, making the carbonyl carbon atom have an increased δ+ charge. Chlorine is also a good leaving group, and leaves as chloride (Cl⁻). There is not much overlap between the chlorine lone-pairs and the carbonyl π* orbital due to a large difference in orbital sizes so resonance forms are not so important. Acyl bromides exist but they are less widely used because they are extremely reactive.

➡ When naming branches from the longest alkyl chain, the C=O carbon is C1.

The naming of acyl chlorides is similar to that for carboxylic acids. For example benzoic acid in the acyl chloride form becomes benzoyl chloride and ethanoic acid, ethanoyl chloride (Figure 7.9). If there are substituents present, the carbonyl carbon is C1.

Figure 7.9 Some acyl chlorides you might encounter.

7.5 ACYL CHLORIDES

How are they prepared?

Acid chlorides are most often prepared from the corresponding carboxylic acid. Most often $SOCl_2$, oxalyl chloride $((ClCO)_2)$ or PCl_5 are used. There are other reagents available, for example Ghosez' reagent (1-chloro-*N*,*N*,2-trimethyl-1-propenylamine), but these are beyond the scope of this book.

How do they react?

Acyl chlorides react with most nucleophiles. They are commonly used to prepare ketones, esters, amides, and acid anhydrides.

Worked example 7.5A

Suggest a mechanism for the following reaction.

$$\text{CH}_3\text{COOH} \xrightarrow{PCl_5} \text{CH}_3\text{COCl} + POCl_3$$

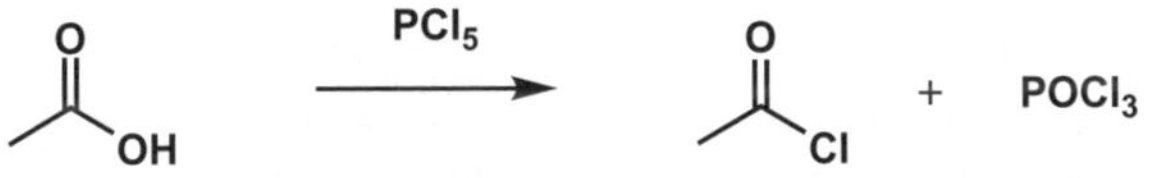

Solution

PCl_5 (phosphorus pentachloride) is a common reagent for this transformation. It has a trigonal bipyramidal structure so the central phosphorus can be easily attacked generating an oxonium intermediate (an oxygen with a positive charge), which can then be attacked by the chloride. Finally, the tetrahedral intermediate falls apart eliminating phosphoryl chloride and the acid chloride. The driving force for this reaction is the formation of a very strong P=O bond.

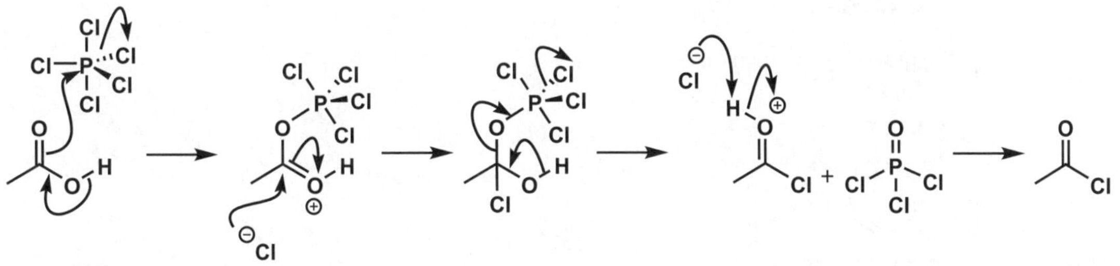

Worked example 7.5B

Suggest a mechanism for the following reaction.

$$\text{2-methylpentanoyl chloride} \xrightarrow{EtMgBr} \text{4-methylhexan-3-one}$$

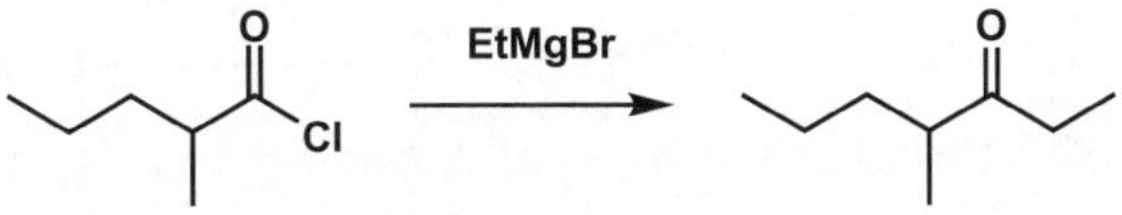

Solution

The Grignard reagent is an excellent nucleophile and will readily attack the acid chloride. We don't have to worry about over-addition to form the tertiary alcohol because the ketone product is much less reactive that the starting material.

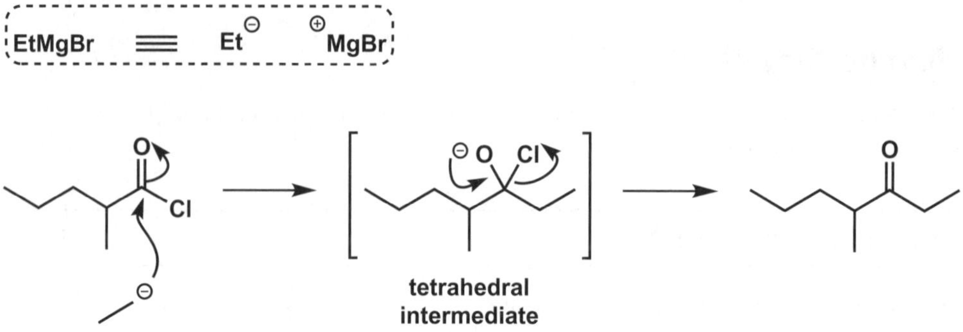
EtMgBr
Et
MgBr
O
Cl
O
Cl
O
tetrahedral intermediate

Question 7.13

Suggest a mechanism for the reaction below.

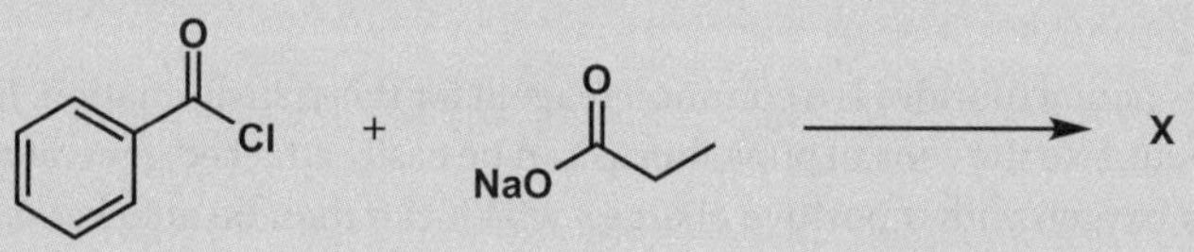
O
Cl
+
O
NaO
X

Question 7.14

Suggest a mechanism for the reaction below.

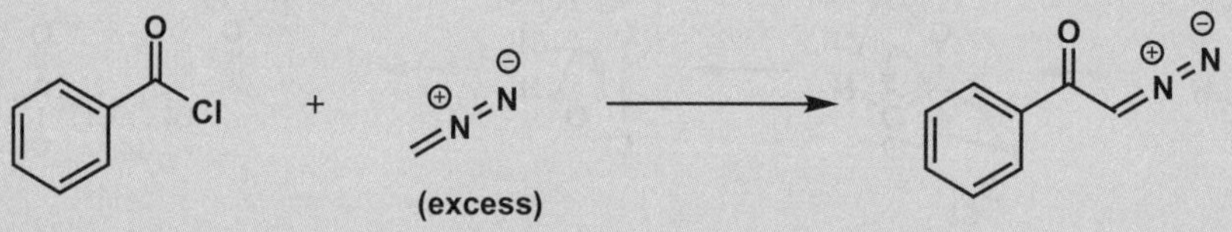
O
Cl
+
N
N
(excess)
O
N
N

Question 7.15

Suggest the most likely product and a mechanism for the reaction below.

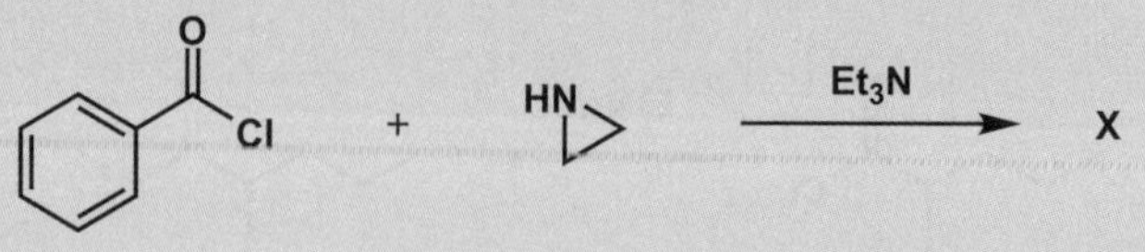
O
Cl
+
HN
Et3N
X

7.6 Esters

What is an ester?

Esters are a class of carbonyl compounds that contain a carbonyl where the central carbonyl carbon is bonded to the two oxygen atoms and another carbon atom (Figure 7.10). Esters are important compounds within the flavourings and fragrances industries because they usually smell pleasant. Butyl butanoate, for example, smells like pear drops. An important thing to note about esters is their nomenclature. Their names are in two parts: the alkyl chain attached to the O is named first, and then the left-hand section (as drawn) up to, and including, the carbonyl group.

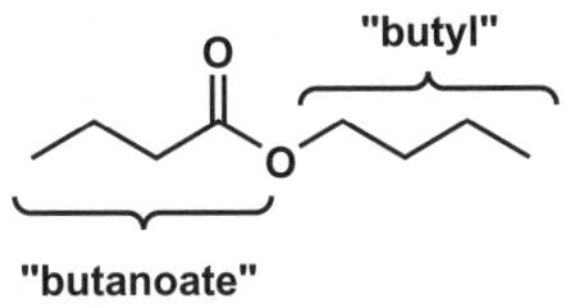

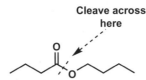

Figure 7.10 Naming of esters.

How are they prepared?

Esters can be made by a variety of methods but the ones you are most likely to encounter are from the reaction of an alcohol with an acid chloride, a carboxylic acid (see section 7.4) or an acid anhydride. When faced with the problem of how to make an ester, the best approach is to imagine cutting the molecule in half across the C—O single bond and then considering which starting materials are most appropriate (Figure 7.11).

Figure 7.11 Cleavage of the ester bond.

Formation from an acid chloride and alcohol usually requires a weak base for two reasons: to remove any HCl by-product which might lead to further undesirable reactions, and to remove the proton from the alcohol once it has attacked the carbonyl group, generating the tetrahedral intermediate (Figure 7.12). This method is generally quite a fast way of making ester compounds but one drawback is the availability and stability of the starting acid chloride; they can be expensive and are hydrolysed to generate a carboxylic acid if not stored correctly.

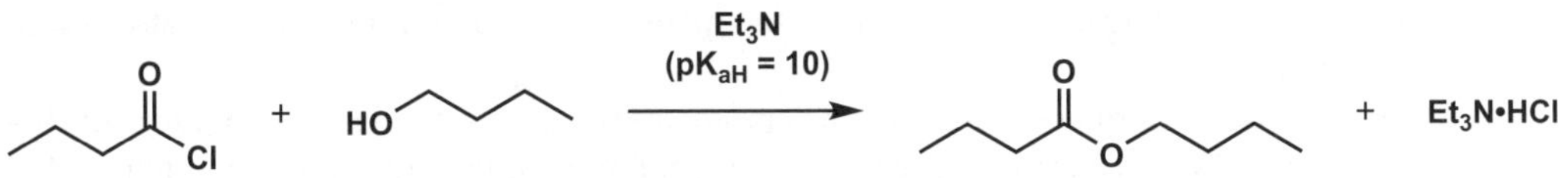

Figure 7.12 Ester formation from an acid chloride and an alcohol.

Preparation from an acid anhydride usually requires slightly more forcing conditions than with an acid chloride because of the reduced reactivity of the anhydride compared to the acid chloride (Figure 7.13). Again, a base is present and is required to remove the proton from the alcohol when it attacks the carbonyl group. When using anhydrides, it is necessary to determine which end the nucleophile will attack otherwise you generate the wrong product!

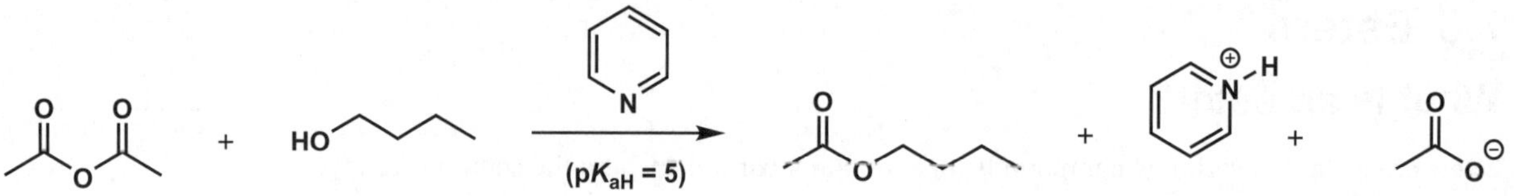

7 CARBONYL CHEMISTRY

Figure 7.13 Ester formation from an acid anhydride and an alcohol.

An alternative route, known as the Fischer synthesis, uses a carboxylic acid in the presence of an acid catalyst (Figure 7.14). The acid activates the carbonyl group by protonation of the carbonly oxygen (for a full mechanism, see Worked example 7.6B). This reaction is in equilibrium therefore it is best to ensure that the water by-product is removed by using a Dean and Stark apparatus to force the equilibrium position to favour the products, or an excess of the alcohol.

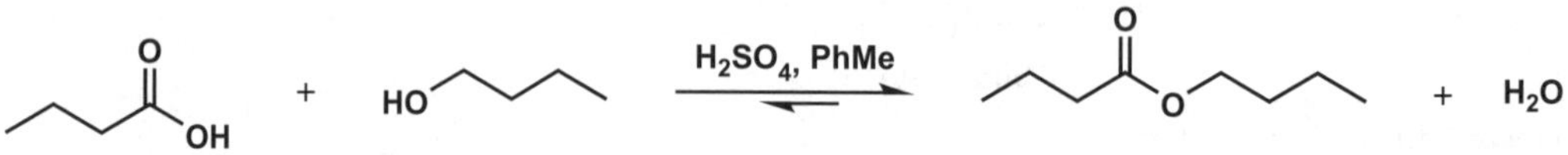

Figure 7.14 Ester formation from a carboxylic acid and an alcohol.

How do they react?

Esters can react to form carboxylic acids (as discussed in section 7.4). They can also be used to generate tertiary alcohols by addition of a good nucleophile, for example a Grignard reagent.

Worked example 7.6A

Suggest a mechanism for the following transformation.

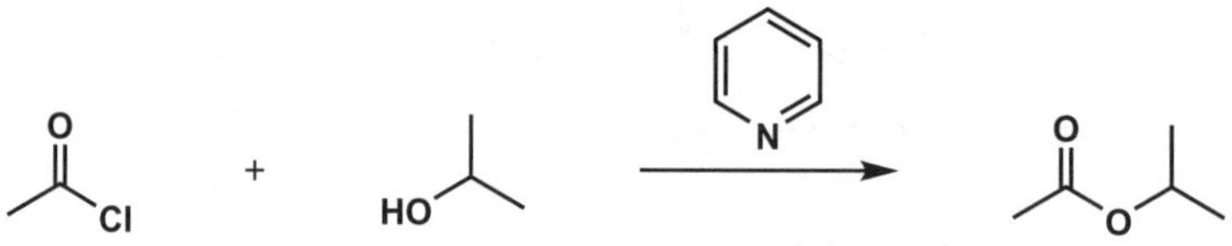

Solution

The reaction shown is a common way of preparing esters. Acid chlorides are extremely reactive therefore the reaction will proceed readily. Pyridine has two roles: one is to neutralize the HCl produced during the reaction by forming a salt (pathway 1), and the other is to activate the acid chloride (pathway 2).

Pyridine is not a strong enough base to deprotonate the alcohol therefore this is not a mechanistic option! As with all reactions where there is addition into a carbonyl group, a tetrahedral intermediate will be formed which, in this case, can eliminate chloride to generate the ester shown.

> If the base used is triethylamine, pathway 2 is less likely to occur due to steric reasons.

7.6 ESTERS

Pathway 1:

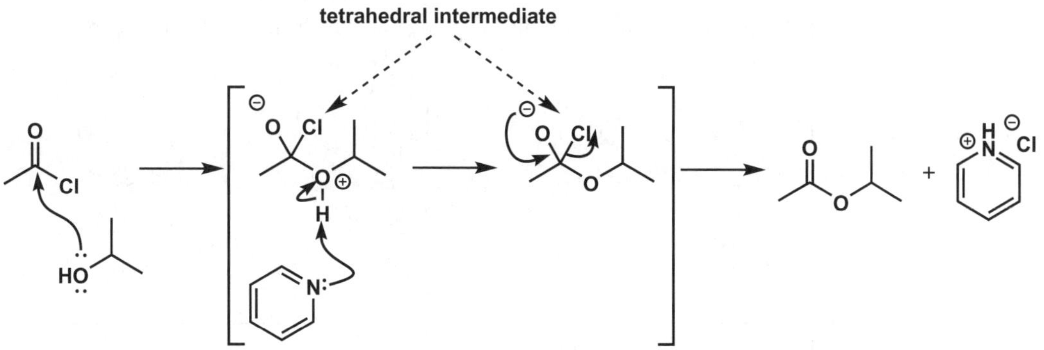

[Mechanism scheme showing acetyl chloride reacting with isopropanol, forming a tetrahedral intermediate which is deprotonated by pyridine, followed by collapse to give isopropyl acetate and pyridinium chloride.]

Pathway 2:

[Mechanism scheme showing pyridine first attacking acetyl chloride to form an activated acyl pyridinium species, which is then attacked by isopropanol to form a tetrahedral intermediate, followed by deprotonation by pyridine and collapse to give isopropyl acetate and pyridinium chloride.]

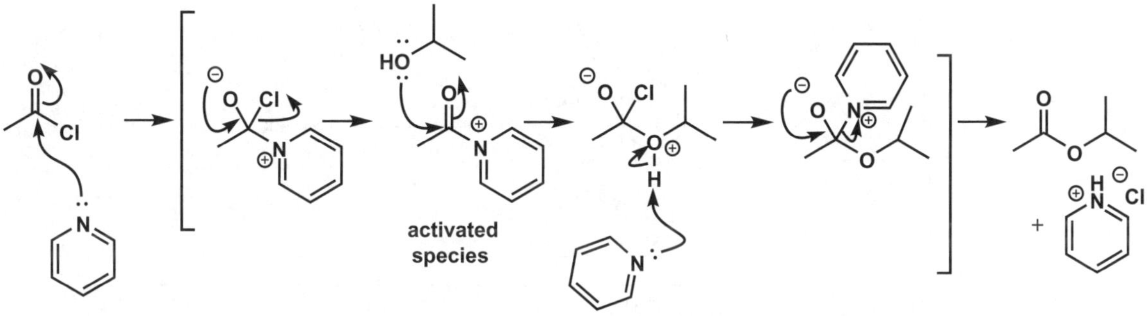

Worked example 7.6B

Suggest a mechanism for the following transformation.

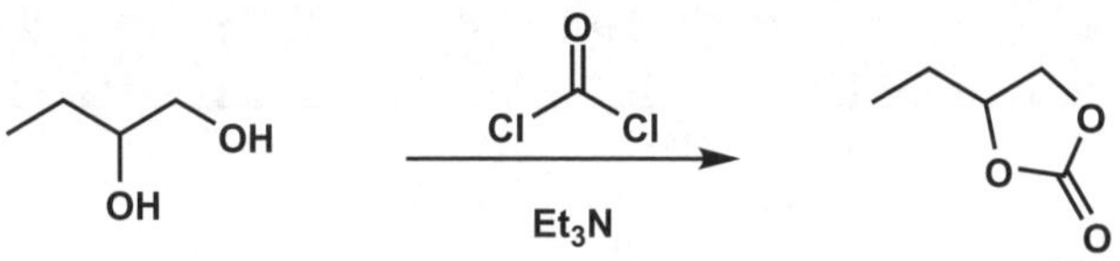

[Scheme: 1,2-butanediol + COCl$_2$, Et$_3$N → 4-ethyl-1,3-dioxolan-2-one (cyclic carbonate)]

Solution

In this reaction the diol has been converted into a carbonate: a carbonyl group with two flanking oxygen atoms. The mechanism for this is similar to that in the previous example; the triethylamine will attack the phosgene (COCl$_2$) to generate an activated intermediate which can then in turn be attacked by the alcohol to form the ester. This process will happen twice. In terms of the order of events, the primary alcohol probably will attack the phosgene first due to steric reasons, then the secondary alcohol will attack to close the ring due to its close proximity to the acyl chloride.

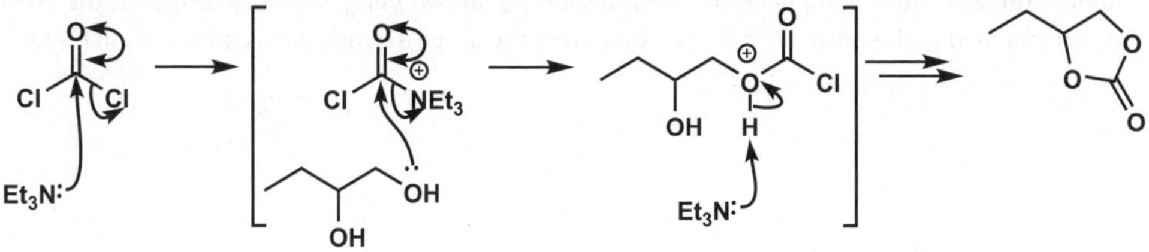

[Mechanism scheme: Et$_3$N attacks phosgene (COCl$_2$) to form an activated acyl-ammonium intermediate, the primary OH of the diol attacks this species, Et$_3$N deprotonates to give the mono-acylated intermediate bearing a free secondary OH, which cyclizes to give the cyclic carbonate product.]

> **Question 7.16**
>
> Suggest a mechanism for the following reaction. What are the roles of pyridine?
>
>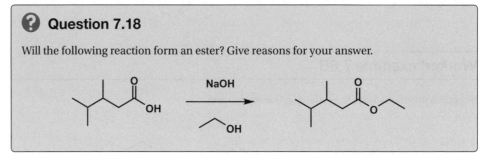

> **Question 7.17**
>
> Acetylsalicylic acid is a pro-drug for aspirin (salicylic acid). Suggest which bond is hydrolysed once it is in the stomach? Suggest a mechanism for this transformation.

> **Question 7.18**
>
> Will the following reaction form an ester? Give reasons for your answer.

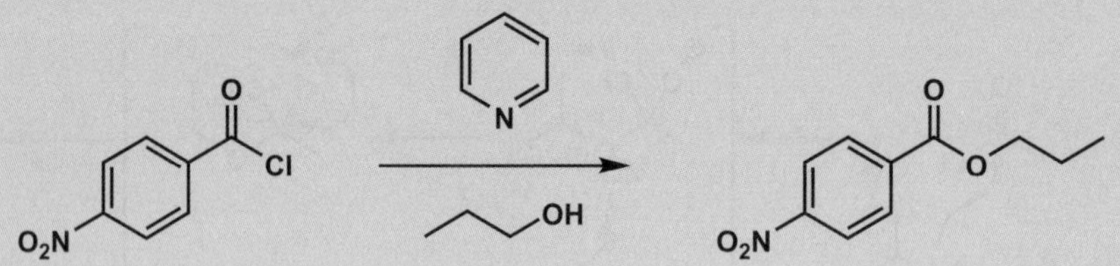

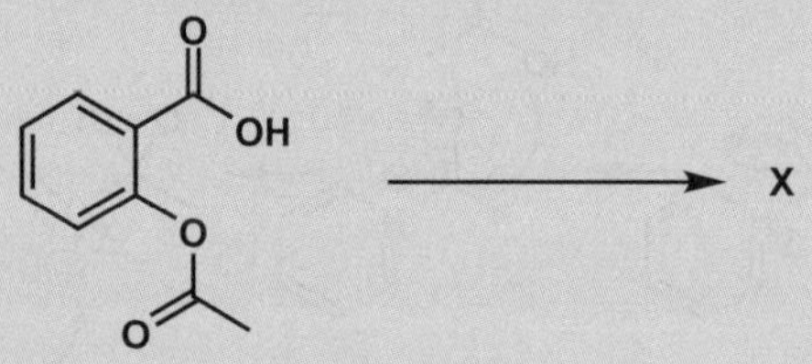

7.7 Amides

What is an amide?

> In chemistry we refer to these compounds as amides but in biochemistry and chemical biology, where multiple amino acids are joined together, these bonds are often referred to as peptides.

Amides are a class of carbonyl compounds that contain a carbonyl group in which the carbonyl carbon is also bonded to a nitrogen and another carbon atom. Amides are important compounds biologically because the amide bond is present in peptides and proteins.

Amides are more resistant to attack than esters because the nitrogen lone pairs can delocalize more effectively into the carbonyl group due to better orbital overlap, and because it is less electronegative than oxygen. The implication of this is that the central carbon atom is less electrophilic and the oxygen atom is more nucleophilic (Figure 7.15). An important thing to note about amides is their nomenclature. Their names are in two parts, similar to an ester: the alkyl chain attached to the nitrogen is named first, then the longest continuous carbon chain including the carbonyl group of the amide.

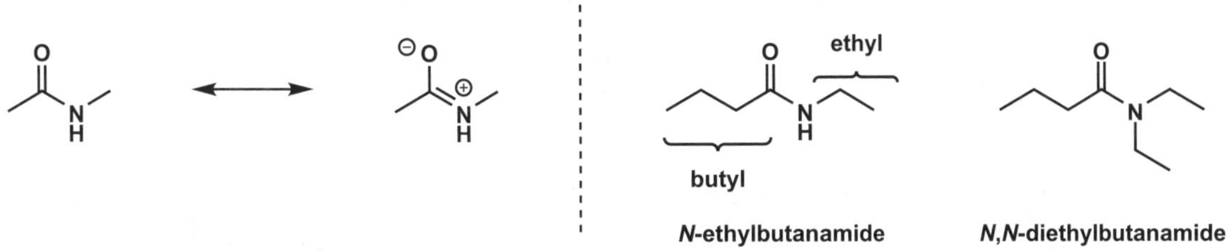

7.7 AMIDES

Figure 7.15 Structure and nomenclature of amides.

How are they prepared?

Amides can be prepared in a number of ways but the ones you are most likely to encounter in your first years of university are reactions between an amine and an ester or an acid chloride. They can also be made by reaction between an amine and a carboxylic acid but this usually requires a coupling agent which is beyond the scope of this book. Mechanistically there is nothing too new with their preparation compared to the esters and acid chlorides discussed previously.

How do they react?

Amides react in much the same way as esters, but often need more forcing conditions due to the increased stability of the amide functionality. They can be hydrolysed to give a carboxylic acid and an amine, and reduced to give an amine. Reaction with Grignard reagents can be sluggish and give poor yields. There is one exception to this however; a Weinreb amide (*N*-methyl-*N*-methoxy amide) can be used to generate ketones from an amide in excellent yields and selectivity.

➔ Weinreb amide

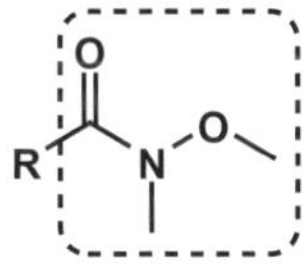

Worked example 7.7A

Suggest a mechanism for the following reaction:

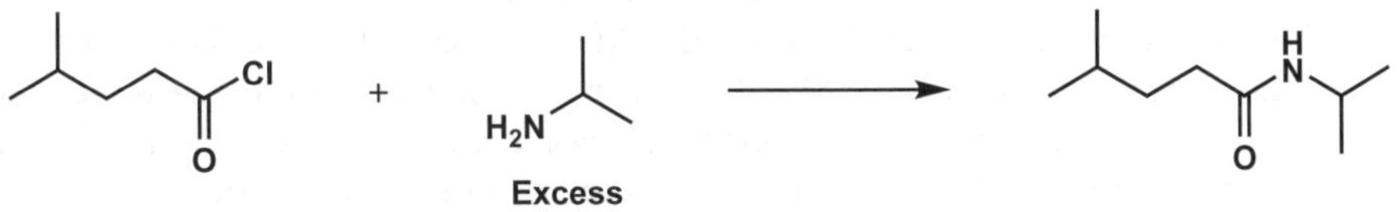

Solution

When looking at this reaction, we can treat it the same as the other carbonyl reactions we have met. The acyl chloride is highly electrophilic at the carbonyl carbon due to the electronegativity of the chlorine, so is highly reactive towards nucleophiles, and chlorine is an excellent leaving group (leaving as Cl⁻). The amine is a good nucleophile because the nitrogen contains a lone-pair which can attack the C=O π^* orbital, generating a tetrahedral intermediate which can then eliminate chloride. The reason that chloride is eliminated and not the amine is because of their respective pK_{aH} values. The pK_{aH} of chloride (i.e. HCl) is −8 but that of the amine (i.e. RNH₂) is approximately 36. This means that the chloride anion is more stable, because it has a lower pK_{aH}, therefore is the preferred leaving group. For this reaction to work the amine must be in excess because one molecule of HCl is released which will make a salt with the amine present.

➔ Remember, pK_{aH} is the PK_a value for the protonated parent species. For a refresher, see Chapter 3, section 3.4.

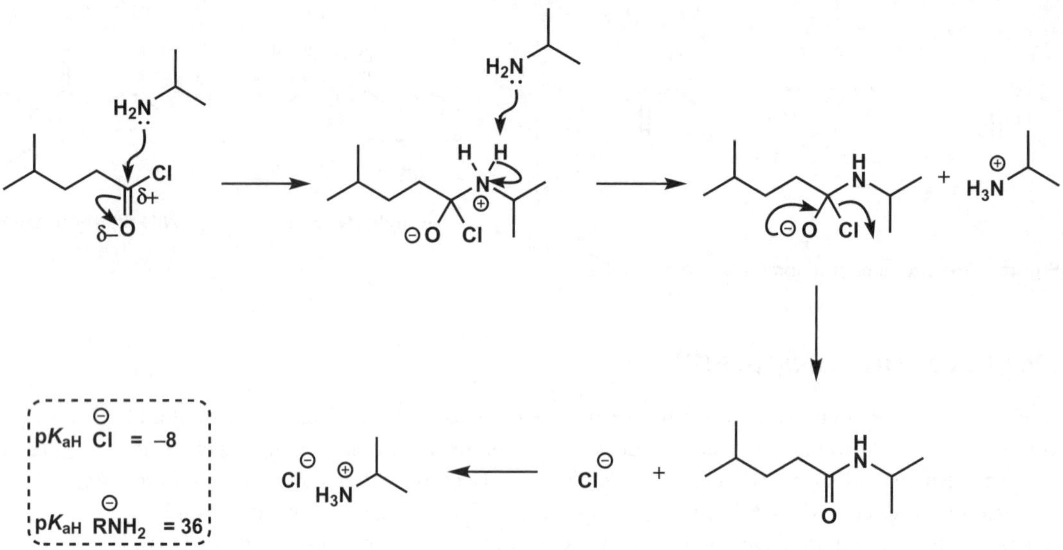

Worked example 7.7B

Suggest a mechanism for the following reaction.

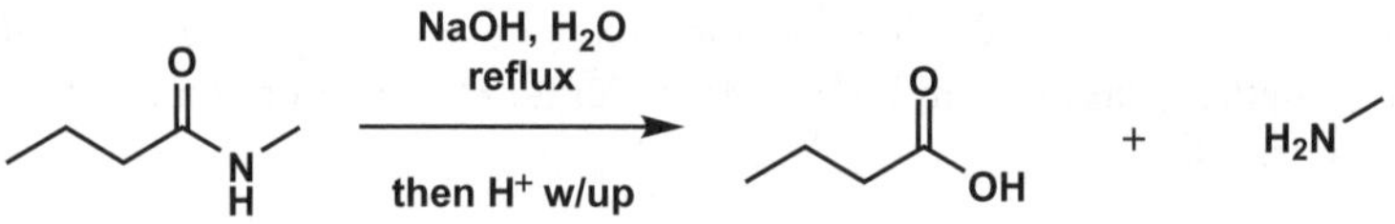

[Reaction scheme: butyramide (N-methyl) + NaOH, H₂O reflux then H⁺ w/up → butyric acid + H₂N-Me]

Solution

Hydrolysis of an amide requires much more forcing conditions than for an ester therefore this reaction must be undertaken at reflux. The hydroxide ion attacks the δ+ carbonyl carbon to generate the tetrahedral intermediate. This then collapses to eliminate the amine as the anion. The pK_a of an amine is approximately 32 and that of water is 15.7, therefore the amine anion formed deprotonates the water present. The newly formed carboxylic acid is deprotonated and therefore cannot react any further due to delocalization of the negative charge, which renders the carbonyl group unreactive towards nucleophiles.

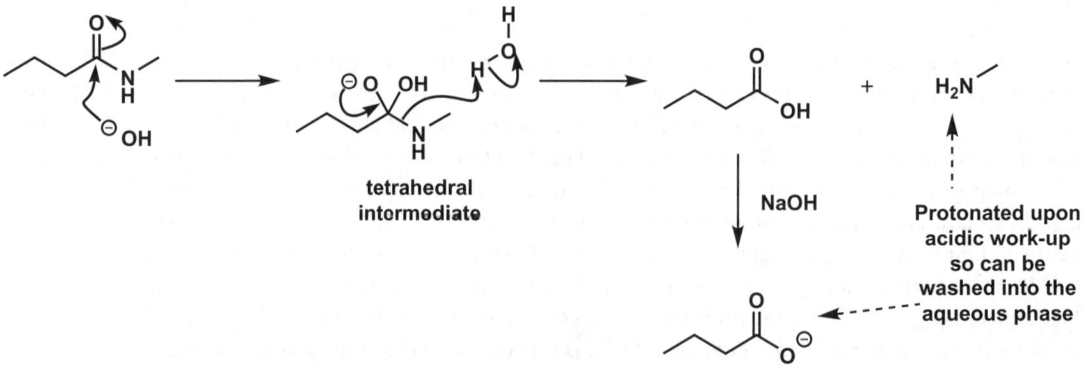

[Mechanism scheme showing: amide + ⁻OH → tetrahedral intermediate → carboxylic acid + H₂N-Me; then NaOH step giving carboxylate. Note: "Protonated upon acidic work-up so can be washed into the aqueous phase"]

Question 7.19

Suggest a mechanism for the reduction below.

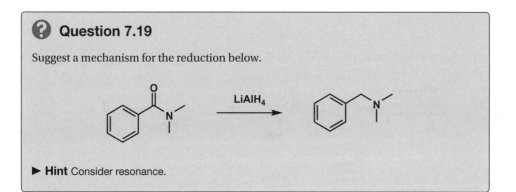

▶ **Hint** Consider resonance.

Question 7.20

Suggest a mechanism for the reaction below.

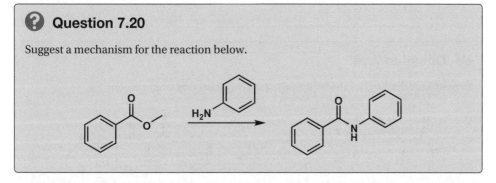

Question 7.21

Suggest the mechanism for conversion of the Weinreb amide shown below to a ketone. How would you make the Weinreb amide?

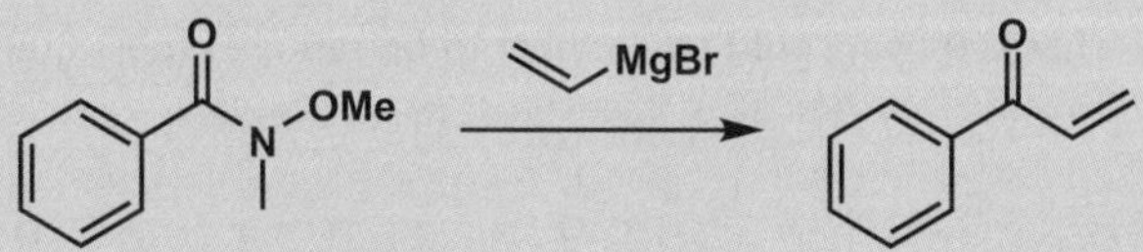

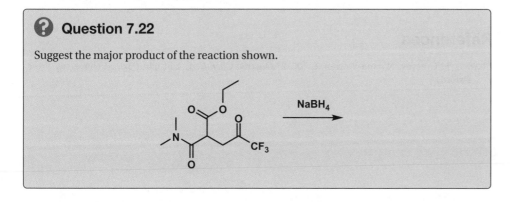

7.8 Synoptic questions

Question 7.22

Suggest the major product of the reaction shown.

7 CARBONYL CHEMISTRY

Question 7.23

Give reagents and conditions for steps A and C. Suggest mechanisms for steps A and C.

Question 7.24

Suggest a mechanism for the hydrolysis of the β-lactam ring in ampicillin.

Question 7.25

Which of the following ketones would you expect to be reduced most quickly by $NaBH_4$ and why?

References

Clayden, J., Greeves, N., and Warren, S. (2012) *Organic Chemistry*, 2nd edn (Oxford University Press, Oxford).

Synoptic questions

S1

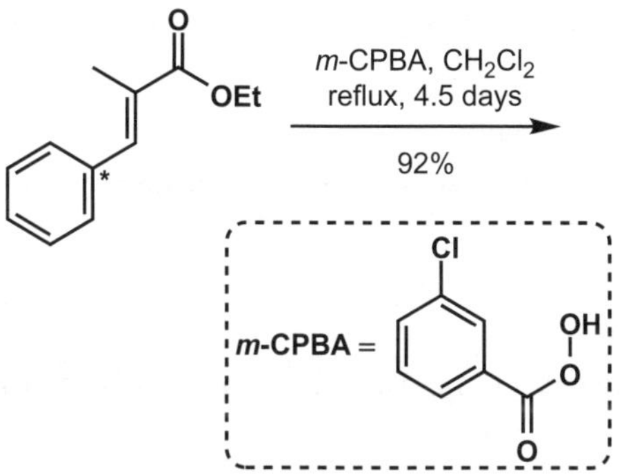

[Structure: ethyl (E)-2-methyl-3-phenylacrylate (cinnamate derivative) with OEt, carbonyl, methyl, and phenyl group; asterisk on the β-carbon]

m-CPBA, CH$_2$Cl$_2$
reflux, 4.5 days
92%
→ X

m-CPBA = 3-chloroperoxybenzoic acid (structure shown with Cl, OH, O–O, C=O)

The following relate to the reaction scheme shown.

(a) Showing your working, is the alkene (E) or (Z)?
(b) What is the hybridization of the C marked with an asterisk?
(c) Predict the product of the reaction with m-CPBA and provide a mechanism.

S2

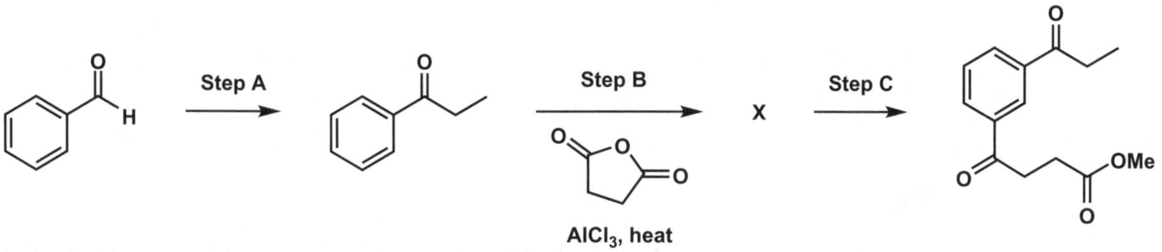

PhCHO →[Step A] PhC(O)Et →[Step B, succinic anhydride, AlCl$_3$, heat] X →[Step C] 1,3-disubstituted benzene with propanoyl group and –CH$_2$CH$_2$C(O)OMe with additional ketone

The following relate to the reaction scheme shown.

(a) Provide reagents, conditions, and a full mechanism for step A.
(b) What is the name of the reaction in step B? Draw a full mechanism for formation of X, fully justifying any regioselectivity.
(c) Give reagents, conditions, and a full mechanism for conversion of X to the product in step C.

S3

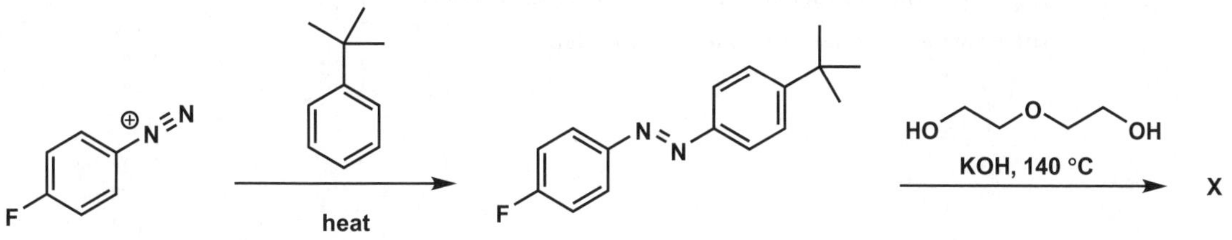

4-F-C$_6$H$_4$–N$_2^+$ + tert-butylbenzene →[heat] 4-F-C$_6$H$_4$–N=N–C$_6$H$_4$-tBu →[HOCH$_2$CH$_2$OCH$_2$CH$_2$OH, KOH, 140 °C] X

The following relate to the reaction scheme shown.

(a) Draw the mechanism for the azo coupling step.

(b) Suggest the structure of X and provide a mechanism for its formation.

(c) Starting from 1-fluoro-4-nitrobenzene, suggest a route for synthesis of the starting diazonium species **A**. (Hint: It will take two steps.)

S4

(2S,3S)-2-bromo-3-methylpentane reacts with lithium diisopropylamide (LDA) to yield two major products.

(a) Draw the structure of (2S,3S)-2-bromo-3-methylpentane.

(b) Predict the structure of the two products, providing mechanisms for each reaction.

S5

Hydroboration–oxidation can be used to form alcohols from alkenes. The process is shown below.

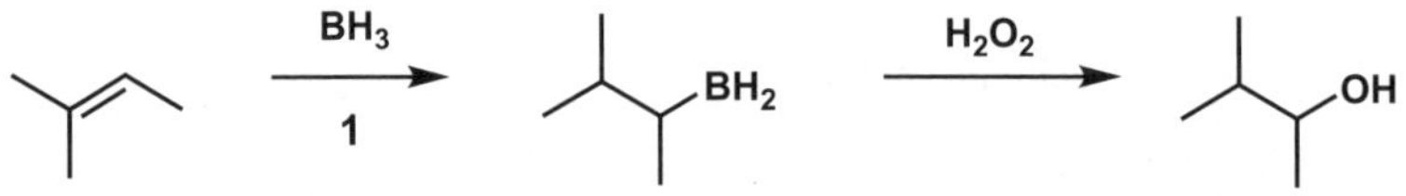

(a) Step 1 is an electrophilic addition. Which reagent is the nucleophile, and which is the electrophile?

(b) Is the molecule synthesized the Markovnikov or anti-Markovnikov product?

(c) Suggest which reagents may be used to synthesize the tertiary alcohol below from this alkene. Provide a mechanism.

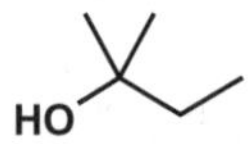

S6

En route to synthesizing diphenhydramine derivatives, you conduct the two-step reaction below.

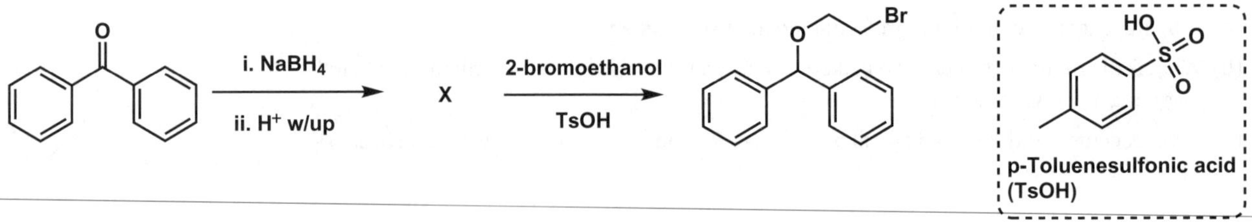

(a) Predict the product from the first step of the reaction.

(b) Provide a mechanism for the second step.

Answers

Final answers to questions posed in the text (where they can be given) are presented here. You can find **full solutions** to every question featured in the book in the *Workbooks in Chemistry* Online Resource Centre. Go to http://www.oxfordtextbooks.co.uk/orc/chemworkbooks/.

CHAPTER 1

Question 1.3
(a) 3-methylpentane
(b) 4-ethyl-5-methylhept-1-ene
(c) Propanoic acid
(d) 3-methylcyclohexan-1-ol
(e) 4-bromopentan-2-ol
(f) 1,2-dichlorobenzene, or *ortho*-dichlorobenzene

Question 1.5
Bond order: 0

Question 1.6
(a) a σ bonding MO and a σ* anti-bonding MO will be formed
(b) a π bonding MO and a π* anti-bonding MO will be formed

Question 1.7
(a) C1, C2: sp^3; C3, C4: sp^2
(b) C1: sp^2; C2: sp^3; C3, C4: sp
(c) C1, C2: sp^2; C3, C4: sp^3
(d) C1: sp; C2: sp^2
(e) Phenyl group at C3 of propyne: $6 \times sp^2$. C3 of propyne: sp^3; C1, C2 of propyne: sp
(f) All C atoms sp^2 hybridized

Question 1.8
(a) O: sp^3 hybridized
(b) O/N: sp^2 hybridized
(c) N: sp hybridized
(d) N: sp^2 hybridized

Question 1.9
(a) 1
(b) 1
(c) 2
(d) 4
(e) 15
(f) 0
(g) 1
(h) 16

Question 1.10
6 DBEs; S(VI) invalidates equation 1.2.

Question 1.11
(a) Nonpolar
(b) Polar
(c) Nonpolar
(d) Polar
(e) Polar

Question 1.12
(a) B
(b) A
(c) A
(d) B

Question 1.13
(a) Aromatic
(b) Non-aromatic
(c) Non-aromatic
(d) Aromatic
(e) Aromatic
(f) Anti-aromatic
(g) Non-aromatic
(h) Aromatic

Question 1.14
(a) Aromatic
(b) Non-aromatic
(c) Aromatic
(d) Aromatic
(e) Non-aromatic
(f) Aromatic

Question 1.16
(a) A
(b) A
(c) B
(d) B

Synoptic question 1.19
(a) N/A
(b) Non-aromatic
(c) N/A

Synoptic question 1.20
'X' is the enolate of 'A'

CHAPTER 2

Question 2.2
(a) Positional isomers
(b) Chain isomers
(c) Functional group isomers
(d) Chain and functional group isomers

Question 2.3
(a) —OH > —NH_2 > —CH_3 > —H
(b) —C≡CH > —C(H)=CH_2 > —CH_3 > —H
(c) —NH_2 > —Et > —CH_3 > —H
(d) —NO_2 > —NMe_2 > —NH_2 > —CN

Question 2.4
(a) *Trans*
(b) *Cis*
(c) *Cis*
(d) *Trans*

Question 2.5
(a) (Z)
(b) (Z)
(c) (E)
(d) (Z)

Question 2.7
(a) (S)
(b) (R)
(c) (R)
(d) (S)
(e) (R)
(f) (S)

Question 2.8
(a) Enantiomers
(b) Not stereoisomers
(c) Meso
(d) Enantiomers
(e) Diastereomers
(f) Diastereomers

Synoptic question 2.9
(a) 1-bromobutane, 2-bromobutane, 1-bromo-2-methylpropane, 2-bromo-2-methylpropane.
(b) (*R*)-2-bromobutane; (*S*)-2-bromobutane

Synoptic question 2.10
(a) Both (*S*)
(b) They are enantiomers
(c) N/A

CHAPTER 4

Question 4.1
(a) E2
(b) E1
(c) E1
(d) E2
(e) E1cB

Question 4.2
4

Appendix 1
Acidity constants

Values of pK_a quoted at 298 K in water except when otherwise noted.

Acid	Formula	pK_a*
Hydriodic acid	HI	−10
Hydrobromic acid	HBr	−9
Hydrochloric acid	HCl	−7
Sulfuric acid	H_2SO_4	−3
Perchloric acid	$HClO_4$	−1.6
Nitric acid	HNO_3	−1.4
Trichloroethanoic acid	CCl_3CO_2H	0.66†
Iodic acid	HIO_3	0.78
Oxalic acid	$(CO_2H)_2$	1.25
Phosphonic acid (phosphorous acid)	H_3PO_3	1.3†
Dichloroethanoic acid	Cl_2CHCO_2H	1.35
Sulfurous acid	H_2SO_3	1.85
Chlorous acid	$HClO_2$	1.94
Hydrogensulfate ion	HSO_4^-	1.99
Phosphoric acid	H_3PO_4	2.16
Chloroethanoic acid	$ClCH_2CO_2H$	2.87
Bromoethanoic acid	$BrCH_2CO_2H$	2.90
Hydrofluoric acid	HF	3.20
Nitrous acid	HNO_2	3.25
Methanoic acid	HCO_2H	3.75
Hydrogenoxalate ion	$HO_2CCO_2^-$	3.81
Benzoic acid	$C_6H_5CO_2H$	4.20
Ethanoic acid	CH_3CO_2H	4.76
Phenylammonium ion	$PhNH_3^+$	4.87
Propanoic acid	$CH_3CH_2CO_2H$	4.87
Pyridinium ion	$C_5H_5NH^+$	5.23
Carbonic acid	H_2CO_3	6.35
Hydrogen sulfide	H_2S	7.05
Hydrogensulfite ion	HSO_3^-	7.2
Dihydrogenphosphate ion	$H_2PO_4^-$	7.21
Hypochlorous acid	HClO (or HOCl)	7.40
Hydrazinium ion	$NH_2NH_3^+$	8.1
Hypobromous acid	HBrO (or HOBr)	8.55
Pentane-2,4-dione	$MeCOCH_2COMe$	9.0
Hydrocyanic acid	HCN	9.21
Ammonium ion	NH_4^+	9.25
Boric acid	H_3BO_3 (or $B(OH)_3$)	9.27†
Trimethylammonium ion	Me_3NH^+	9.80
Silicic acid	H_4SiO_4	9.9‡

(continued)

Acid	Formula	pK_a*
Phenol	C_6H_5OH	9.99
Hydrogencarbonate ion	HCO_3^-	10.33
Ethylammonium ion	$EtNH_3^+$	10.65
Methylammonium ion	$MeNH_3^+$	10.66
Triethylammonium ion	Et_3NH^+	10.75
Hydrogen peroxide	H_2O_2	11.62
Hydrogenphosphate ion	HPO_4^{2-}	12.32
Water	H_2O	14.00
Methanol	MeOH	15.5
Hydrogensulfide ion	HS^-	19
Propan-2-one	MeCOMe	20
Ethyne	C_2H_2	25
Hydrogen	H_2	35
Ammonia	NH_3	38
Benzene	C_6H_6	43
Ethene	C_2H_4	44
Ethane	C_2H_6	50

* Values below −2 and above 18 are approximations.
† 293 K.
‡ 303 K.

Sources

Haynes, W.M. (ed.) (2015–16). *CRC handbook of chemistry and physics*, 96th edn. CRC Press, Boca Raton, Florida.
Smith, M.B. and March, J. (2007). *March's advanced organic chemistry: reactions, mechanisms, and structure*, 6th edn. Wiley-Interscience, New York.

Appendix 2
Electronegativity values for common elements

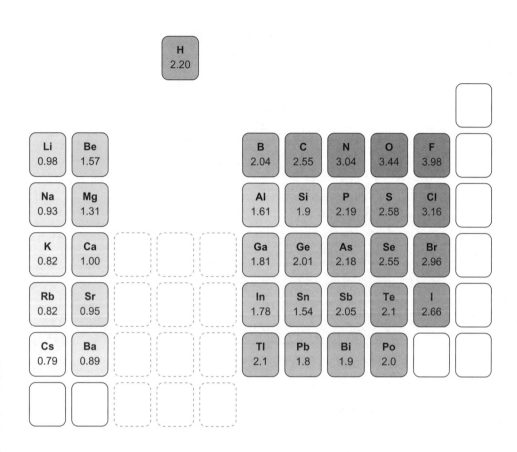

Appendix 3

Common functional groups in decreasing order of seniority, according to IUPAC

Functional Group	Structure	Prefix	Suffix
Radicals	R•	–	-yl
Anions	R⊖	–	-ide
Cations	R⊕	–	-ylium
Carboxylic acids	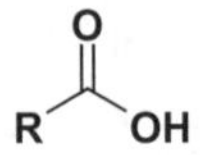 R-C(=O)-OH	Carboxy-	-oic acid
Esters	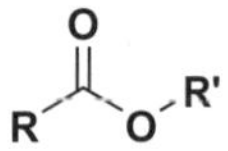 R-C(=O)-O-R'	Alkoxy-oxo-	-oate
Acid halides	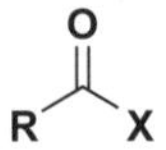 R-C(=O)-X, X = Cl, Br, I	Halo-oxo-	-oyl halide
Amides	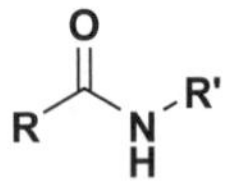 R-C(=O)-NH-R'	Amino-oxo-	-amide
Nitriles	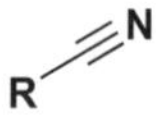 R-C≡N	Cyano-	-nitrile
Aldehydes	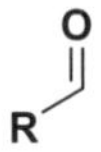 R-CHO	Oxo-	-al
Ketones	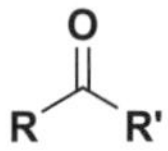 R-C(=O)-R'	Oxo-	-one
Alcohols	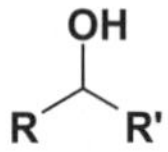 R-CH(OH)-R'	Hydroxy-	-ol
Thiols	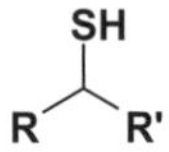 R-CH(SH)-R'	Sulfanyl-	-thiol
Amines	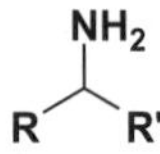 R-CH(NH$_2$)-R'	Amino-	-amine
Imines	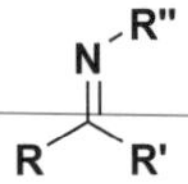 R-C(=N-R'')-R'	Imino-	-imine
Alkene	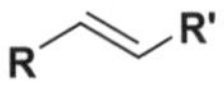 R-CH=CH-R'	–	-ene
Alkyne	R-C≡C-R'	–	-yne

Index

Page numbers in bold type refer to sidenotes.
Page numbers italics refer to figures.

1-chloropentane 34
2-chloroacetic acid 19, 20
2,2-dichloroacetic acid 19–20
3-methylbutanal 33
3-methylphenol 25–6

A

acetic acid 19
acetone 4, 26-7, 29–30
acetylacetone 26–7
acetylsalicylic acid 106
acid anhydrides 87
   preparation of esters 103–4
acid/acyl chlorides 87, 100–2
   preparation of esters 103
acidity constants 115–16
addition reactions
   electrophilic addition 67–70
   halogen addition 68–9
   Michael addition 31
alcohols 5
aldehydes 5, *87*, 89, 90
alkyl halides 5
amides *87*, 90, 106–9
amines **57**
amino acids 7
ampicillin 88, 89, 110
anti-aromatic molecules **21**
anti-bonding orbitals 8
aromatic molecules/rings 21–3, **26**
   azo coupling 82–4
   electrophilic substitution 72–8
   nucleophilic substitution 79–82
arrow notation **18**
aspirin 88, 89
atomic orbitals 8
Aufbau principle **10**
azo coupling 82–4

B

benzene 16
benzoic acid 100
bimolecular elimination 62
bond angles 1, **3**
bond orders 8–9
bonding orbitals 8
borane 94, 96
borohydride 94
boron trichloride 47

Brønsted acids/bases 49, 50
Bürgi–Dunitz angle 91
butanone 28–9

C

Cahn–Ingold–Prelog (CIP) system 35, 37–8
capsaicin 16–17
carbonyl groups 87–91
   reactions with nucleophiles 91–4
   reactions with reducing agents 94–7
   see also acid/acyl chlorides, amides, carboxylic acids, esters
carboxylic acids 5, *87*, 97–100
catenation **1**
chain isomers 32, 34
chiral molecules 39–41
   see also optical isomers
cis/trans isomers 35, 36–8
condensed formulae **1**
configurational isomers 32, 35
conformational isomers 32
constitutional isomers 32–4
CORN rules **39**
cyclohexene 71
cyclooctatetraene 22–3
cyclopentadiene 22
cyclopropanone 90

D

di-*iso*-butylaluminium hydride (DIBAlH) 94, 96
diastereomers 40, 42–3
diazonium salts 82
dichloromethane 18, **68**
dipole moments 18
directing groups 74–9
double bond equivalents 15–17

E

E/Z isomers 36
E1 elimination 62–3, 65
E1cB elimination 62, 63
E2 elimination 62, 63
electronegativity values 117
electrophiles 47–8
electrophilic addition 67–70
electrophilic substitution 72–8
elimination reactions 62–5
   bimolecular elimination 62
   E1 elimination 62–3, 65

   E1cB elimination 62, 63
   E2 elimination 62, 63
   unimolecular elimination 62
enantiomers 40–1, 42–3, 53
esters *87*, 90, 91, 103–6
ethylene glycol 69–70

F

Fischer esterification/synthesis 99, 104
formaldehyde **4**, 18–19
Friedel–Crafts mechanisms 73
functional group isomers 32, 34
functional groups 5–7, 8, 118
   geminal functional groups **68**
   vicinal functional groups **68**
furan 22

G

geminal functional groups **68**
Grignard reagents **93**, 102, 104

H

halogen addition 68–9
halohydrins **68**
highest occupied molecular orbitals (HOMOs) 8, 88
Hückel's rule 21, 22
Hund's rule **10**
hybrid orbitals 11–14
hydride-containing reducing agents 94–5
hyperconjugation 52

I

in-phase orbitals 8
inductive effect 19–20, **74**
isoleucine 7
isomers 32–43
   chain isomers 32, 34
   cis/trans isomers 35, 36–8
   configurational isomers 32, 35
   conformational isomers 32
   constitutional isomers 32–4
   diastereomers 40, 42–3
   E/Z isomers 36
   enantiomers 40–1, 42–3, 53
   functional group isomers 32, 34
   meso isomers 40, 42–3
   optical isomers 35, 39–43
   positional isomers 32, 34
   (*R*)/(*S*) isomers 39, 40–1
   stereoisomers 32, 35

isotopes 35
IUPAC nomenclature/notation 4–7, **18**, 118

K

Kekulé's structure **16**
keto-enol tautomerism 28
ketones 5, 29, *87*, 89

L

L-propargylglycine 13
lactams *87*, **89**
lactones *87*
leaving groups 57
Lewis acids/bases 49, 50
Lewis structures **1**
linalool 8
lithium aluminium hydride 94, 95, 96
locants **4**
lowest unoccupied molecular orbitals (LUMOs) 8, 88

M

Markovnikov's rule 67
meso isomers 40, 42–3
mesomeric effect 25, **74**
methyl tert-butyl ether (MTBE) 2
Michael addition 31
molecular formulae **1**
molecular orbitals 8–14
  anti-bonding orbitals 8
  bonding orbitals 8
  highest occupied molecular orbitals 8, 88
  hybrid orbitals 11–14
  in-phase orbitals 8
  lowest unoccupied molecular orbitals 8, 88
  out-of-phase orbitals 8
molecular polarity 18–20
muscone 88, 89

N

Newman projections **62**
nitro group **75**
nomenclature
  see IUPAC nomenclature
nucleophiles 47–8
  reactions with carbonyl groups 91–4
nucleophilic substitution 51–6
  aromatic molecules 79–82
  S_N1 nucleophilic substitution 51–3
  S_N2 nucleophilic substitution 51, 54–6, **63**

O

optical isomers (chirality) 35, 39–43
orbitals
  see atomic orbitals, molecular orbitals
out-of-phase orbitals 8

P

Pauli exclusion principle **10**
penguinone 30
phosgene 105
phosphorous pentachloride 101
pK_a values 19–20, 26–7, 56–60
polar bonds 18
polarity
  see molecular polarity
positional isomers 32, 34
propadiene 14
pseudoephedrine 46
pyridine 104

R

$(R)/(S)$ isomers 39, 40–1
reducing agents 94–7
regioselectivity 67–8
resonance 24–7

S

skeletal formulae 1–2
S_N1 nucleophilic substitution 51–3
S_N2 nucleophilic substitution 51, 54–6, **63**
sodium bromide 48
sodium cyanide 92
sodium cyanoborohydride 94, 95
sodium hydroxide 98
sp^2 hybridized carbon atoms 14
stereocentres 35, **40**
stereoisomers 32, 35
structural formulae **1**
substitution reactions
  electrophilic substitution 72–8
  nucleophilic substitution 51–6
sulfate anion 24

T

tautomerism 28–30
  keto-enol tautomerism 28
tert-butyl group 2
tetrachloromethane 18
toluene **4**
tosyl chloride 17
tosylate/tosyl group **48**
triethylamine 104, 105
trivial names **4**

U

unimolecular elimination 62

V

vicinal functional groups **68**

W

Wheland intermediates 72, 73
Weinreb amide 107, 109

Z

Zaitsev's rule 64